AGRICULTURE ISSUES AND POLICIES

FARMING AND FARMLAND IN THE UNITED STATES

CHANGES AND TRENDS

AGRICULTURE ISSUES AND POLICIES

Additional books in this series can be found on Nova's website under the Series tab.

Additional E-books in this series can be found on Nova's website under the E-book tab.

AGRICULTURE ISSUES AND POLICIES

FARMING AND FARMLAND IN THE UNITED STATES

CHANGES AND TRENDS

CALVIN IGLEHART
AND
GRANT J. ZSOFKA
EDITORS

nova publishers
New York

For permission to use material from this book please contact us:
Telephone 631-231-7269; Fax 631-231-8175
Web Site: http://www.novapublishers.com

Additional color graphics may be available in the e-book version of this book.

Library of Congress Cataloging-in-Publication Data

ISBN: 978-1-62257-907-5

Published by Nova Science Publishers, Inc. † New York

CONTENTS

PREFACE

This book examines the changes and trends in farming and farmland in the United States. Innovations in farm organization, business arrangements, and production practices have allowed farmers to produce more with less. Fewer labor hours and less land are used today than 30 years ago, and practices such as the use of genetically engineered seeds and no-till have dampened increases in machinery, fuel, and pesticide use. Likely aided by the increased use of risk management tools such as contracts and crop insurance, U.S. agricultural productivity has increased by nearly 50 percent since 1982. Future innovations will be necessary to maintain, or boost, current productivity gains in order to meet the growing global demands that will be placed upon U.S. agriculture.

Chapter 1 – Innovations in farm organization, business arrangements, and production practices have allowed farmers to produce more with less. Fewer labor hours and less land are used today than 30 years ago, and practices such as the use of genetically engineered seeds and no-till have dampened increases in machinery, fuel, and pesticide use. Likely aided by the increased use of risk management tools such as contracts and crop insurance, U.S. agricultural productivity has increased by nearly 50 percent since 1982. Future innovations will be necessary to maintain, or boost, current productivity gains in order to meet the growing global demands that will be placed upon U.S. agriculture.

Chapter 2 – Because farm real estate represents much of the value of U.S. farm sector assets, large swings in farmland values can affect the financial well-being of agricultural producers. This chapter examines both macroeconomic (interest rates, prices of alternative investments) and parcel-specific (soil quality, government payments, proximity to urban areas) factors that affect farmland values. In the last few years, U.S. farmland values have been supported by strong farm earnings, which have helped the farm sector in many regions to withstand the residential housing downturn. Historically low interest rates are likely a significant contributor to farming's current ability to support higher land values. About 40 percent of U.S. farmland has been rented over the last 25 years. Non-operators (landowners who do not themselves farm) owned 29 percent of land in farms in 2007, though that proportion has declined since 1992.

In: Farming and Farmland in the United States ISBN: 978-1-62257-907-5
Editors: C. Iglehart and G. J. Zsofka

Chapter 1

THE CHANGING ORGANIZATION OF U.S. FARMING*

Erik J. O'Donoghue, Robert A. Hoppe, David E. Banker, Robert Ebel, Keith Fuglie, Penni Korb, Michael Livingston, Cynthia Nickerson and Carmen Sandretto

ABSTRACT

Innovations in farm organization, business arrangements, and production practices have allowed farmers to produce more with less. Fewer labor hours and less land are used today than 30 years ago, and practices such as the use of genetically engineered seeds and no-till have dampened increases in machinery, fuel, and pesticide use. Likely aided by the increased use of risk management tools such as contracts and crop insurance, U.S. agricultural productivity has increased by nearly 50 percent since 1982. Future innovations will be necessary to maintain, or boost, current productivity gains in order to meet the growing global demands that will be placed upon U.S. agriculture.

---

* This is an edited, reformatted and augmented version of United States Department of Agriculture, Economic Research Service Economic Information Bulletin No. 88, dated December 2011.

Photo: USDA/ARS and Shutterstock.

SUMMARY

What Is the Issue?

If global population and energy demands grow as expected, and if prices continue to fluctuate, current productivity gains may not keep pace with the increasing demands placed upon U.S. agriculture. Recognizing where to devote limited resources to ensure that continuing innovation takes place to meet these future demands requires understanding how production, managerial practices, business arrangements, and productivity influence each other.

What Did the Study Find?

Over the past 30 years, a series of inter-related changes in input use, business arrangements, farm structure, and production practices combined to expand output without increasing the use of total inputs. Moreover, by allowing farmers to increase U.S. agricultural production through increased productivity instead of expanded land and chemical use, many of these innovations helped to limit the impact of agricultural production on the environment.

- Use of two major inputs, land and labor, has decreased over time. From 1982 to 2007, land used in agriculture dropped from 54 to 51 percent of total U.S. land area, while farming used 30 percent less hired labor and 40 percent less operator labor. Meanwhile, new technologies (such as precision agriculture)—often requiring new or advanced management techniques—have been increasingly adopted, particularly by farmers.
- Farmers have altered how they manage their risk, including a heavier reliance on contracting (the value of production under contract increased roughly 10 percentage points between 1991 and 2007) and a shift of production to farms organized as partnerships and corporations (from 34 percent of all farm product sales in 1982 to 43 percent by 2007), allowing risks to be spread over a wider set of stakeholders. Federal crop insurance has also become a major risk management tool (farmers insured 100 million acres in 1989; by 2007, over 270 million acres were insured).
- Larger farms receive the bulk of commodity payments while most conservation payments accrue to smaller farms. Overall, payments are smaller yet make up a larger share of gross cash farm income for smaller farms, which often rely heavily on off-farm income, while larger farms receive larger payments that make up a much smaller share of their gross cash farm income. Over the past 3 decades, government policies have shifted from a concentration on supply management to focus on income support, while a growing emphasis has been placed on environmental concerns—most recently on working land programs.
- Despite declines in the use of land and labor, agricultural productivity has maintained a linear growth pattern. Driven by the increased use of technology, production practices have changed, such as the increased use of no-till (from 5 percent of all planted acres in 1989 to 23 percent by 2004) and the reduced use of pesticides on many crops. Many of these changes have also lowered labor requirements, which have allowed some farms to increase the size of their operations. Although production has shifted dramatically to larger farms over the past 25 years, 97 percent of all farms remain family farms, generating 82 percent of the total value of U.S. agricultural production.

How Was the Study Conducted?

This study drew upon data from various sources. USDA's National Agricultural Statistics Service (NASS) and the Economic Research Service (ERS) jointly design and administer multiple surveys annually, known collectively as USDA's Agricultural Resource Management Survey (ARMS), which covers U.S. farming operations and their operators in the 48 contiguous States (see www. ers.usda.gov/Briefing/ARMS/). The ARMS Phase III 1996-2008 surveys provided detailed information on farm business organization, contracting, operator demographics, government payments, and production practices.

The census of agriculture provides comprehensive historical data on consolidation and specialization trends. Data on trends in government payments came from the *U.S. and State Farm Income* series (the sector accounts) maintained by ERS. Various NASS publications, such as the *Acreage* reports, provided additional data on genetically engineered crop adoption and other crop production practices. National Resources Inventory (NRI) data provided land use estimates, and administrative data concerning program payments are from the agencies issuing them.

INTRODUCTION

While U.S. farms generated less than 1 percent of total U.S. gross domestic product (GDP) in 2007, they remain an important segment of the U.S. and global economy. Although U.S. agriculture has maintained productivity increases while reducing resource use, expected world population growth and increasing energy demands could soon outstrip its capacity to meet these needs. An open question is whether future productivity gains can match previous trends without harming the natural resources farmers require as crucial inputs or jeopardizing the quality of U.S. soils, air, and water supplies.

Productivity increases require more efficient use of existing resources, and land is one of the most important inputs to farming. Over the past 3 decades, gains in productivity have allowed for increased production on less cropland. Most changes in cropland levels are due to farmers and landlords shifting land between cropland and grazing or forest land uses, while some land leaves agriculture altogether due to urban development pressures.

Farmers face many sources of risk, which can cause productivity to suffer. They rely heavily upon their inputs, some of which (precipitation, sunshine) cannot be controlled. Farmers also face risks from pests and diseases, which can destroy crops and livestock. In addition, global population growth and drives to generate renewable energy could contribute to high and fluctuating oilseed and grain prices, putting increasing pressure on the resources used by farmers.

Policy Issues

The problems that face the agricultural sector vary over time due to changes in technology, local and global market conditions, and consumer concerns. For example, consumers demand safe yet affordable food, while also professing concern for both small farms and family farms. Often, new situations engender the introduction of new policies—a process that may create an inefficient patchwork of government programs.

Farming Evolves

Examining changes in farm organization, production practices, Government policies, and productivity over the past three decades enables informed thinking about the innovations that have allowed the agricultural sector to successfully adapt to changing market conditions in the past. Over 1982- 2007, use of many farm inputs, including land and labor, remained constant or declined, while productivity grew through improvements in technology, increased use of some chemical inputs, and changes in production practices like increases in the use of conservation tillage. Greater technology adoption, however, has heightened farm operators' need for education and has encouraged farmers to increase the size of their operations. In turn, larger farms have led farmers to rely more on contracting and corporate forms of organization, spreading risks over a wider set of stakeholders. Federal crop insurance and income support programs also help farmers absorb risk. Production has shifted to larger farms that are more likely to be profitable, better able to adopt new technologies, and more likely to use production or marketing contracts. In contrast, the number of smaller commercial farms (annual sales $10,000-$250,000) has declined, while noncommercial farms, with sales below $10,000, largely exist independently of the farm economy. However, despite the continued shift of production to larger farms, farming has remained primarily a family business. The family-farm share of the farm count remained between 97 and 99 percent from 1988 through 2007.

Data, Current and Historic

This report draws upon data from various sources. USDA's National Agricultural Statistics Service (NASS) and the Economic Research Service (ERS) jointly design and administer multiple annual surveys, known collectively as the Agricultural Resource Management Survey (ARMS). ARMS covers U.S. farms and the farmers who operate them in the 48 contiguous States (see www.ers.usda.gov/Briefing/ARMS/). The 2007 ARMS Phase III (which coincides with the most current census of agriculture) provides detailed information on farm business organization, contracting, operator demographics, government payments, and production practices. For long-term consolidation and specialization trends, however, we rely largely on the census of agriculture, which dates back to 1840 (ARMS has been in use only since 1996). Data on government payments come from administrative data maintained by the agencies that make the payments, while data on trends in government payments back to 1982 come from the *U.S. and State Farm Income* series maintained by ERS. Another ERS data series, *Agricultural Productivity in the United States*, provides data on farm productivity. Various NASS publications, such as the *Acreage* reports, provide additional data on genetically engineered crop adoption and other crop production practices. For land use estimates, this report draws upon the ERS Major Land Use series and the National Resources Inventory (NRI) from USDA's Natural Resources Conservation Service. Finally, we summarize results from the literature to provide insights into why various changes took place.

The 1982-2007 Period

We chose the 1982-2007 period because it is the most recent 25-year period covered by the census of agriculture. From 1982 until 2002, there was no clear trend in prices for farm products, as measured by the producer price index (PPI) for farm products. After 2002, however, prices have trended upward, at least through 2010. Higher prices in recent years mean that farmers face a different economic environment—with strong price signals— which appears to be continuing. Not all data sources cover the entire 1982-2007 period. In particular, we follow contracting, farm operators' education, and selected business arrangements from 1991 to 2007 using ARMS and its predecessor, the Farm Costs and Returns Survey (FCRS). Also, our examination of farming stops at the farm gate, precluding topics such as the effects of international trade or changing consumer preferences on farm organization.

The Farm Definition

The USDA defines a farm as any place that produced and sold—or *normally* would have produced and sold—at least $1,000 of agricultural products (including government payments) during a given year (O'Donoghue et al., 2009b). If a place does not have sales of at least $1,000, a "point system" assigns values for acres of various crops and head of livestock to estimate normal sales (where 1 point equals 1 dollar in estimated sales). "Point farms" are farms with sales less than $1,000 but points worth $1,000 or more.

Point farms' share of the farm count has grown over time, increasing from 11 percent of all farms in the 1982 Census of Agriculture to 31 percent in 2007 (Hoppe et al., 2010). Some of this increase, however, is due to methodological changes in the census of agriculture (see box, "Counting Farms: Methodological Changes," p. 4). One of these changes—weighting census estimates for undercoverage—is particularly important, since it increased the count of point farms by 50 percent.[1] In addition, NASS has increased its efforts to contact more small farms when conducting the census of agriculture.

By 2007, farms with sales less than $10,000—including point farms—made up 60 percent of all U.S. farms, up from 43 percent in 1982 (measuring sales in constant 2007 dollars). Farms with sales less than $10,000 are more likely to be rural residences than farm businesses. Thus, this report, when appropriate, sorts farms by sales class so that information about the large number of point farms and other farms with sales less than $10,000 does not mask the characteristics of more commercially oriented farms.

Counting Farms: Methodological Changes

Three methodological changes introduced in recent censuses of agriculture have increased the count of U.S. small farms:

1) Data for 1997 through 2007 were adjusted for undercoverage (farms that the census missed), predominantly focusing on farms with sales less than $2,500 (USDA, NASS, 2004). Prior to that time, gross adjustment factors were published. However, these factors were not applied to the published estimates or the estimates in computer files.

2) The 2007 Census of Agriculture counted more farms in the lowest sales categories, due to extensive mailing list building efforts and increased area sampling (USDA, NASS, 2010a). In addition, commodity prices were relatively high in 2007, which means that more establishments were able to meet the $1,000 minimum sales level necessary to be counted as a farm.

3) Beginning in 1997, the census asked the respondent for each Indian reservation to report the number of American Indians operating farms or ranches on the reservation (USDA, NASS, 1999, p. VIII). Prior to 1997, the census counted only one operator per reservation.

The last item in the list obviously increased the count of American Indian operators, most of whom have small farms. The first and second changes, however, increased estimates of small farms in general as well as counts of small farm operators who are women, members of a racial minority, or Hispanic. The changes outlined above improved the quality of census data while complicating the analysis of long-term trends.

While the aforementioned changes took place explicitly, a fourth, more subtle complication involves the definition of a farm. An entity qualifies as a farm if it has annual sales (or normally would have annual sales) of $1,000 worth of agricultural commodities. This definition was first implemented in the 1970s, and has not changed since. Nominal price increases, therefore, implicitly lead to an increasing number of point, and very small farms over time.

MAJOR INPUTS

Labor and land both figure prominently as major inputs into farming. Both labor and land—particularly cropland—have trended downward over time, with labor declining more rapidly than land. While aging farmers make up the bulk of operators, those who run the larger, more productive operations tend to be younger. The majority of agricultural land appears to stay in agriculture over time, and most of the changes in cropland occur as transitions to other agricultural uses, such as grassland, pastureland, and rangeland.

Labor

Farm operators and their families provide much of the labor used in farming—particularly on smaller farms—as well as management. As farm size increases, however, the ability of the operator's family to meet the labor requirements of the farm business diminishes. These farm operators may use hired or contract farm labor, the use of which increases sharply with sales (table 1.1). Fifteen percent of farms with annual sales less than $10,000 used hired/contract labor in 2007, versus 88 percent of farms with sales of $1 million or more.

The person who acts as the employer determines the difference between hired and contract labor. Farm operations pay hired workers directly, and the worker is a farm employee. In contrast, farm operations pay labor contractors (or crew leaders) for the performance of specific tasks by contract laborers.

Amount and Sources of Labor

The amount of labor used per farm grows with farm size, increasing from a median of 0.7 annual person equivalents (1 person working 40 hours per week for 50 weeks) for the smallest farms to 5.6 for the largest (fig. 1.1).

The operator and spouse provide the bulk of the labor on smaller farms—averaging at least three-fifths of total labor hours until sales reach $500,000. Even in the $500,000-$999,999 sales class, the operator and spouse remain an important source of labor, accounting for roughly the same share of labor hours as hired/contract labor. Once sales reach $1 million, however, hired and contract labor account for 89 percent of labor hours.

Table 1.1. Farms reporting hired or contract labor, 2007

Sales class	Hired or contract labor	Hired labor	Contact labor
		Percent	
Less than $10,000	14.9	8.8	7.1
$10,000-$99,999	34.1	21.6	16.1
$100,000-$249,999	47.3	36.3	18.8
$250,000-$499,999	63.7	51.3	24.0
$500,000-$999,999	70.3	58.5	26.1
$1,000,000 or more	88.3	80.7	32.4
All farms	27.4	18.9	12.0

Source: USDA, Economic Research Service and National Agricultural Statistics Service, 2007 Agricultural Resource Management Survey, Phase III.

Hired and operator labor has trended downward in recent years, even as farm output has increased (fig. 1.2). By 2007, farming used 30 percent less hired labor and 40 percent less operator (and unpaid) labor than in 1982, while farm output was 35 percent higher than in 1982.[1] A recent ERS study found that labor hours devoted to farming declined 2.1 percent per year between 1981 and 2004, while labor productivity increased by 3.7 percent per year. About two-thirds of the growth in output per hour of work came from growth in total factor productivity, or technological change, such as biotechnology, improved animal husbandry, improvements in machinery and chemicals, and more efficient farm size/organization (Fuglie et al., 2007). The U.S. agricultural research system, which includes both public and private-sector research, is a major driver of growth in total factor productivity (Fuglie and Heisey, 2007).

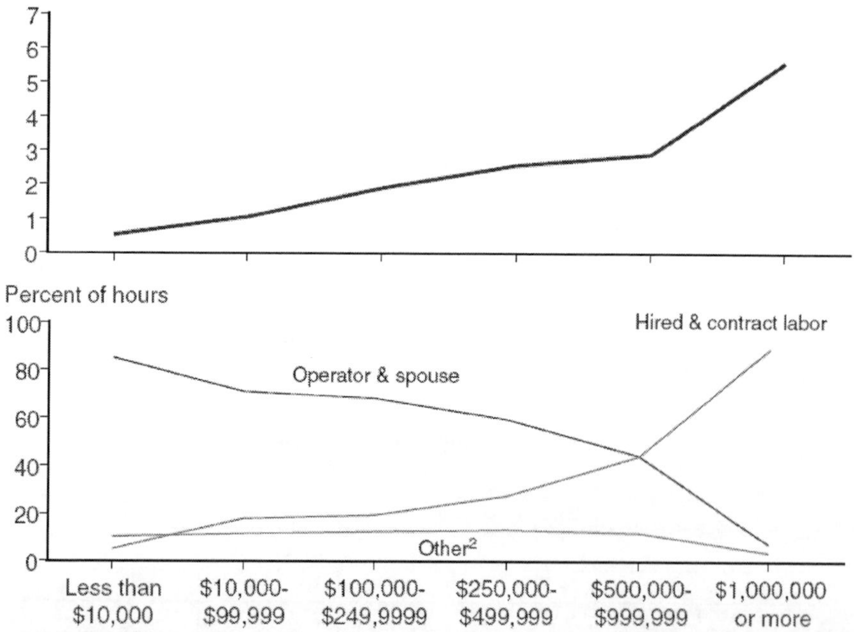

[1]One annual person equivalent equals 2,000 hours or 50 weeks per year times 40 hours per week.

[2]Includes other operators and unpaid workers.

Source: USDA, Economic Research Service and National Agricultural Statistics Service, 2007 Agricultural Resource Management Survey, Phase III.

Figure 1.1. Hours and sources of farm labor by sales class, 2007 Median annual person equivalents per farm.[1]

Farm Operators

Farm operators provide management on farms of all sizes and a significant share of the labor on farms with sales below $1 million. While most farmers are White men—83 percent in 2007—their share has decreased by roughly 10 percentage points since 1982, reflecting recent growth in the number of minority and women operators (see box, "Women and Minority Operators Increase Their Shares," p. 8).

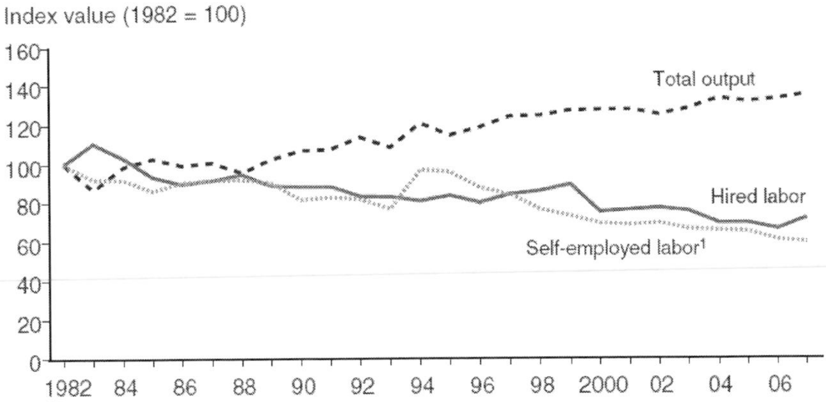

[1]Operators and unpaid workers.

Source: USDA, Economic Research Service, Indices of farm output, input and total factor productivity (http://www.ers.usda.gov/Data/AgProductivity/#datafiles).

Figure 1.2. Indices of total agricultural output, hired labor, and self-employed labor, 1982 to 2007.

Age

By 2007, roughly 30 percent of principal farm operators were at least 65 years old (fig. 1.3), classified as "older operators" here, compared with only 8 percent of self-employed workers in non-agricultural industries (U.S. Dept. of Labor, 2008). In contrast, the youngest farmers, those less than 35 years old, made up roughly 5 percent of principal operators, down from 16 percent in 1982. The downward trend in the youngest farmers reflects farm consolidation, the presence of multiple generations of operators on some farms, and the capital-intensive nature of farming. For example, land prices and startup capital requirements can make it difficult for beginning farmers (especially young farmers with limited credit history) to purchase or rent land.

Why So Many Older Operators?

The advanced age of farm operators is understandable given the low sales requirements to qualify as a farm. Most operators live on their farm, and farmers can phase out of farming over a decade or more while easily retaining enough land or livestock to produce $1,000 in sales. Improved health and advances in farm equipment also allow operators to farm later in life than in past generations (Mishra et al., 2005).

Older operators' share of farms declines with sales class, reflecting their gradual withdrawal from farming (table 1.2). For example, older operators accounted for more than 30 percent of farms with sales up to $99,999, but only 17 percent of farms with sales greater than $500,000. In contrast, operators between 45 and 55 years old accounted for the largest share (34 or 35 percent) of farms with sales greater than $500,000.

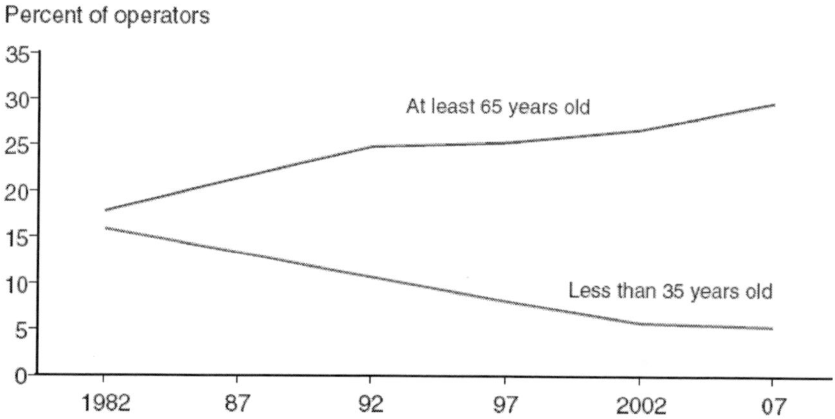

Source: USDA, Economic Research Service, compiled from census of agriculture data.

Figure 1.3. Principal farm operators by age class, 1982 to 2007.

Women and Minority Farmers Increase Their Shares

Between 1982 and 2007, the number of U.S. farm operators stabilized at roughly 2 million. Nonetheless, the number of non-White operators grew by two-thirds, women operators more than doubled, and Hispanic operators more than tripled. Not all the increase, however, came from new entrants. Some of the increase reflects changes in census methodology designed to provide better counts of small and minority-run farms. By 2007, women operated more than 300,000 farms, representing nearly 14 percent of U.S. farms, up from 5 percent in 1982. Most of the farms operated by women are very small. Approximately 78 percent had sales less than $10,000 in 2007, compared with 57 percent of farms operated by men. Still, nearly 2,000 women operated farms with sales of at least $1 million in 2007.

Among minority farms, operations run by American Indians/Alaska Natives or Hispanics grew the fastest, increasing their shares by at least a percentage point (see figure).

Overall, 6.5 percent of U.S. farms in 2007 were run by minority principal operators, up from 3 percent in 1982.[1] Minorities, with the exception of Asians, tend to operate very small farms. A particularly large share of American Indians/Alaska Natives and Blacks operated very small farms—77 and 80 percent, respectively— compared with 59 percent of Whites.

Only 43 percent of Asian farmers operate farms with sales less than $10,000. The share of Asians with million-dollar farms (8 percent) is much higher than for any other group, including Whites (2 percent). The higher sales of Asian farms reflect their specialization in high-value crops (vegetables, fruits, tree nuts, and nursery/ greenhouse products). The majority of Asian operators live in California and Hawaii, two States that specialize in these crops.

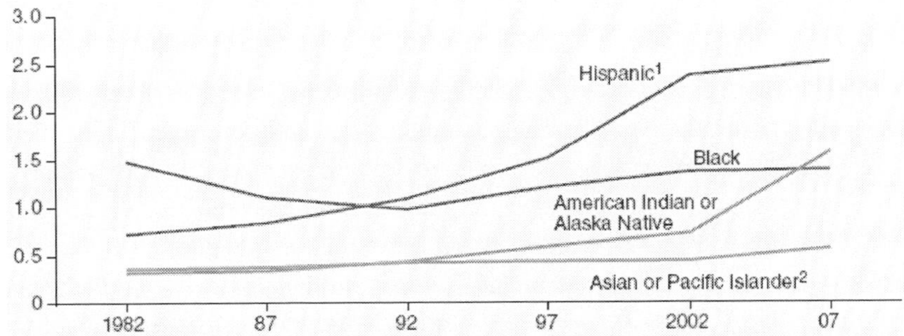

Note: The graph excludes "other races" used prior to 2002 and "more than one race" used in 2002 and 2007.

[1]Hispanics may be of any race. Nonwhite Hispanics, therefore, are counted as Hispanic and as a racial minority in this graph.

[2]Two categories used in 2002 and 2007—"Asian" and "Hawaiian and Pacific

---

[1] Minority operators include any non-Whites and Hispanics. Hispanics, however, may be of any race. Therefore, the percentage of U.S. farms operated by racial or Hispanic minorities—3 percent in 1982 and 6.5 percent in 2007—was adjusted to avoid counting nonwhite Hispanics twice.

Islander"—were combined to be consistent with earlier years.
Source: USDA, Economic Research Service, compiled from census of agriculture data.

Selected minority principal farm operators, 1982 to 2007 Percent of all principal operators.

Table 1.2. Distribution of farms by sales class and age of principal operator, 2007

Sales class	Less than 35 years	Age of principal operator 35 to 65 years	65 years or more
		Percent of sales class	
All principal operators	5.4	64.9	29.7
Less than $10,000	4.7	64.6	30.7
$10,000 to $99,999	6.1	60.5	33.4
$100,000 to $249,999	7.8	67.5	24.7
$250,000 to $499,999	6.8	74.0	19.2
$500,000 to $999,999	5.4	77.6	17.0
$1,000,000 or more	5.0	78.2	16.8

Source: USDA, Economic Research Service, compiled from USDA, National Agricultural Statistics Service, 2007 Census of Agriculture.

Does Age Matter?

The 30 percent of U.S. farmers who were at least 65 in 2007 raises concerns about their impending retirement from agriculture and replenishment by younger farmers. However, the eventual exit of these older principal operators is not as ominous as it first appears (Hoppe and Banker, 2010). Most of the assets of older operators are on retirement or residential/ lifestyle farms that produce only 2 percent of the Nation's farm output. A large share of the land owned by these operators is already rented out or enrolled in land-retirement programs. In addition, some commercially sized farms with older principal operators are multiple-generation farms, with at least 20 years separating the oldest and youngest operators and a replacement operator in the wings.

Education

Farm operators historically graduated from high school at lower rates than the general public, but that gap had largely closed by the late 1980s (Bellamy, 1992). By 2007, the share of farm operators receiving a high school diploma

(90 percent) exceeded the graduation rate for all U.S. households (87 percent) (table 1.3). College degrees remain more common for all U.S. householders than for farm operators, though nearly 30 percent of the operators of million-dollar farms had college degrees in 2007, the same share as for all U.S. householders. The educational gap between operators of million-dollar farms and smaller farms has declined since the early 1980s (fig. 1.4). The increase in college education in the lowest sales class may reflect higher educational levels required for off-farm work.

Table 1.3. Educational attainment, principal operators by sales class and all U.S. householders, 2007

Educational attainment	Principal operators by size of farm: Less than $10,000	$10,000-$99,999	$100,000-$249,999	$250,000-$499,999	$500,000-$999,999	$1,000,000 or more	All principal operators	All U.S. **House-holders**[1]
	Percent of group							
Less than high school	10.2	9.9	10.3	6.0	5.9	4.9	9.7	12.8
Completed high school	41.5	42.2	42.6	41.4	40.8	37.0	41.6	29.4
Some college	25.4	22.5	25.0	27.5	29.5	28.6	24.9	27.3
College graduate	22.9	25.5	22.2	25.1	23.8	29.6	23.8	30.4

[1] Householders age 25 or older.

Source: USDA, Economic Research Service and National Agricultural Statistics Service, 2007 Resource Management Survey for farm operators. U.S. Census Bureau, Current Population Survey for all U.S. households.

Off-Farm Work...

By 2007, 40 percent of U.S. farmers spent at least 200 days per year working off-farm (table 1.4), up from 35 percent in 1982. Off-farm work is most common among the operators of farms with sales less than $10,000, and a substantial increase in very small farms—from 43 percent of all farms in 1982 to 60 percent in 2007—helps account for the increase in off-farm work reported by farm operators.

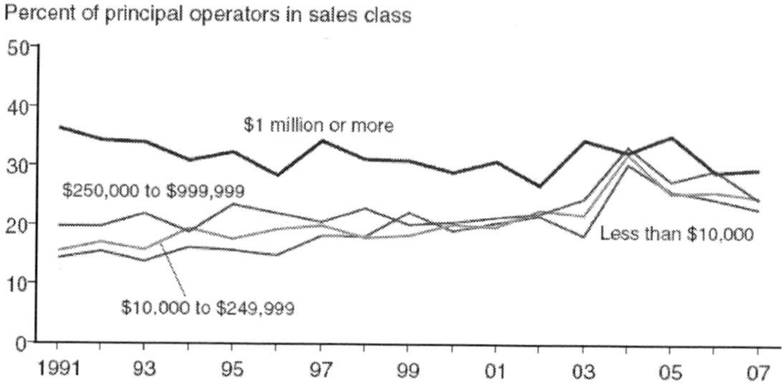

Notes: Consistent educational data for farm operators are available only from 1991 forward. Sales classes are expressed in 2007 dollars, using the Producer Price Index for farm products to adjust for price changes.

Source: USDA, Economic Research Service and National Agricultural Statistics Service, 1991-1995 Farm Costs and Returns Survey and 1996-2007 Agricultural Resource Management Survey.

Figure 1.4. Farm operators who graduated from college, by sales class, 1991 to 2007 Percent of principal operators in sales class.

...and Income

Farm households' off-farm work is reflected in the level and sources of their income. Median income for farm households with sales less than $100,000 was just above the $47,300 average for all U.S. households in 2007, largely due to off-farm earnings (fig. 1.5). About three-quarters of all off-farm income is from wage and salary jobs or self-employment. Older operators also rely on unearned income—such as Social Security, pensions, and interest/dividends.

Nevertheless, farming is an important source of income for households operating farms with sales exceeding $100,000, and it accounts for most household income once sales exceed $250,000. In addition, farm households with sales of at least $250,000 had a larger median income in 2007 (ranging from $92,500 to $202,100 depending on sales class) than did U.S. households with a self-employed head ($75,500).

Agricultural Land

Farming's contribution toward the Nation's gross domestic product (GDP)— averaging roughly 1.3 percent of GDP over the past 10 years—belies the fact that the sector uses a significant portion of the Nation's resources to generate agricultural goods.

Table 1.4. Farms reporting hired or contract labor, 2007

Sales class of farm operated	Reporting off-farm work	Days per year		
		1-99 days	100-199 days	200 days or more
		Percent of sales class		
All principal operators	64.7	16.1	8.9	39.7
Less than $10,000	71.0	17.2	8.9	44.9
$10,000 to $99,999	65.3	15.3	10.8	39.2
$100,000 to $249,999	48.7	14.7	8.4	25.5
$250,000 to $499,999	38.9	14.0	5.8	19.1
$500,000 to $999,999	33.4	11.3	4.1	17.9
$1,000,000 or more	31.0	8.5	3.3	19.1

Source: USDA, Economic Research, compiled from USDA, National Agricultural Statistics Service, 2007 Census of Agriculture.

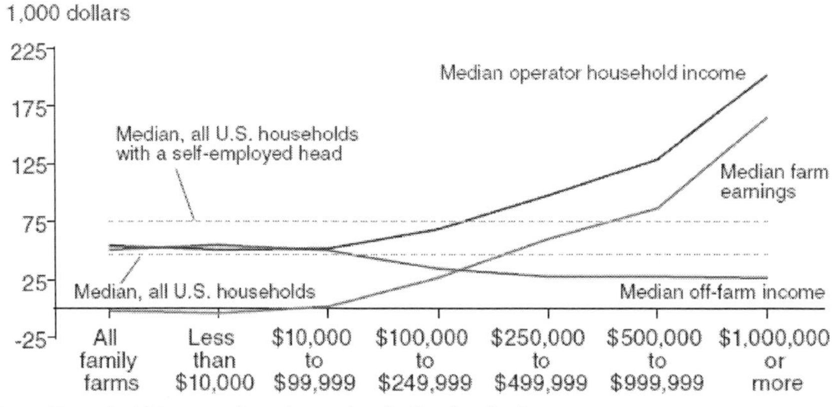

Note: Household income is estimated only for family farms.

Source: USDA, Economic Research Service and National Agricultural Statistics Service, 2007 Agricultural Resource Management Survey, Phase III, for farm households. Federal Reserve Board, 2007 Survey of Consumer Finances for all U.S. households and U.S. households with a self-employed head (Bucks et al., 2009).

Figure 1.5. Median operator household income by source and sales class, 2007 1,000 dollars.

Land is one of the most important of these resources, and land used for agricultural purposes totaled 1.163 billion acres in 2007, representing about 52 percent of total U.S. land area (table 1.5). Of the land in an agricultural use, most has historically been in grassland pasture and range, followed by cropland, grazed forest land, and "special uses," or land in farmsteads and farm roads (see box, "Land Use Definitions," p. 13).

Between 1982 and 2007, the total amount of land in agricultural uses declined by about 5 percent, or 68 million acres, led by decreases in cropland (13 percent) and grazed forest land (18 percent). In 2007, cropland reached its lowest level since ERS began developing consistent estimates of major land uses in 1949.

In contrast, land in grassland pasture and range increased slightly, while farmsteads and farm roads (special uses) rose from 8 to 12 million acres over 1982-2007. The 4-million-acre net increase combines an increase in housing and infrastructure to support farming among commercially oriented farms, a decline from absorbing abandoned farmsteads into agricultural uses, and an increase in farmsteads as the number of small farms with sales less than $1,000 has increased.

Shifts in Agricultural Land Use

While the total amount of land in agriculture has remained fairly constant, land has shifted among agricultural uses due to commodity price changes, evolving agricultural policies, and, more recently, bioenergy policies. The National Resources Inventory (NRI) land-use transition matrices show that the 2002-07 period accounted for some of the largest gross flows into and out of cropland since 1982. Between 2002 and 2007, a total of 14 million acres exited cropland uses while 4 million acres transitioned into cropland used for crops, for a net loss of 10 million acres.[2]

Most of the changes in cropland uses occurred due to transitions between cropland uses, pasture/rangeland, and forest—between 2002 and 2007, a net 6.5 million acres of cropland was lost due to transitions between these three land use categories (USDA, NRCS, 2009). Roughly 9.3 million acres (65 percent of the total land exiting cropland) moved into a pasture or rangeland use while 700,000 acres transitioned to forest.

Of land shifting into cropland uses, nearly 3.4 million acres (75 percent of the total land entering) had been in pasture or rangeland uses in 2002 (fig. 6).

Table 1.5. Agricultural uses of land, 1982 and 2007

	1982	2007	Change
	Million acres		*Percent*
Grassland, pasture, and range	597	614	2.8
Cropland	469	408	-13.0
Grazed forest land	157	129	-17.8
Special uses (farmsteads, farm roads)	8	12	50.0
Total	1,231	1,163	-5.5

Note: Includes all 50 States. See box, "Land Definitions" for explanations of their land-use categories.

Source: For 1982 see citations in Frey and Hexem 1985. For 2007 data see citations in Nickerson et al. 2011.

Land Use Definitions

The Economic Research Service maintains the Major Land Use (MLU) time series (Nickerson et al, 2011). It is the only consistent account of all U.S. land uses, both public and private. Definitions used in the data are outlined below. The series is based on data from the census of agriculture, the Bureau of Land Management, the Forest Service, and other Federal and State agencies.

Agricultural land—Includes all relevant land uses, regardless of ownership. This includes more land than the concept of "land in farms" used by USDA's National Agricultural Statistics Service. Land in farms consists primarily of agricultural land used for crops, pasture, or grazing. Land in farms also includes woodland and wasteland not being used for agricultural purposes (crop cultivation or grazing) as long as it was part of a farmer's total operation, as well as idled cropland (land used for cover crops or soil improvement, and land in conservation programs such as the Conservation Reserve Program and the Wetlands Reserve Program). While agricultural land includes these land uses, it also includes land owned by the government and used for grazing under government permits on a per-animal head basis, as well as an estimate of grazing on forested Federal and non-Federal land that is not in farms.

Cropland—Includes cropland used for crops, cropland used for pasture, and idle cropland. The estimate of total cropland is based on the census of agriculture, which has been conducted by USDA's National Agricultural Statistics Service roughly every 5 years since 1945.

Cropland used for crops—Includes harvested cropland, crop failure, cultivated summer fallow. Cropland used for crops represented 82 percent of total cropland in 2007.

Cropland pasture—Pasture that is generally considered to be in long-term crop rotation. It also includes land used for pasture before crops reach maturity. Cropland pasture and permanent grassland pasture have not always been clearly distinguished in surveys. Nine percent of total cropland was in cropland pasture in 2007.

Idle cropland—Includes land in cover and soil-improvement crops and cropland on which no crops were planted. Cropland enrolled in the Federal Conservation Reserve Program (CRP) and Wetland Reserve Program (WRP) is included in idle cropland. Idle cropland made up 9 percent of total cropland in 2007.

Grassland pasture and range—Includes all open land used primarily for pasture and grazing. The MLU estimates include pasture and range in farms, estimates of private grazing land not in farms, and public, nonforested grazing land.

Grazed forest land—Includes brush-grown pasture, arid woodlands, and other areas within forested areas that have grass or forage. This land also includes woodland pasture in farms as well as estimates of forested grazing not in farms. For many States, the estimates include significant areas grazed only sporadically.

Special uses—Includes farmsteads and farm roads.

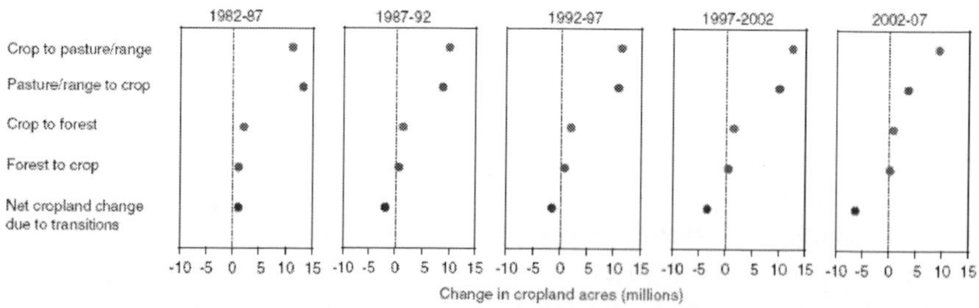

Note: Total cropland: 419.6 million acres (1982); 357.0 million acres (2007).

Source: USDA, Economic Research Service analysis of National Resources Inventory data (USDA, NRCS 2009).

Figure 1.6. Cropland transitions to/from selected land cover/use categories, 1982-2007.

While movements between major land uses happen annually, they constitute a relatively small proportion of the total agricultural land base. Between 2002 and 2007, 96 to 99 percent of agricultural land remained in its pre-existing use (table 1.6). Larger land use transitions took place over the long run, with land in forest changing much more slowly than land used for crops and pasture/range.

Table 1.6. Percent of land remaining in same use, 1982-2007

Land use	1982-87	1987-92	Census year 1992-97	1997-02	2002-07	1982 to 2007
			Percent			
Crop	93	92	95	93	96	78
Pasture/range	95	96	95	96	98	86
Forest	98	98	98	98	99	92

Source: USDA, Economic Research Service analysis of National Resources Inventory (USDA, NRCS, 2009).

Land Used for Crops Declined in Every Region from 1982 to 2007

Land use patterns can vary greatly by region, reflecting differences in soils, climate, topography, and patterns of population settlement. Cropland used for crops, a category including land that is actively cropped or in short-term summer fallow (excluding idle cropland and cropland pasture—see box, "Land Use Definitions," p. 13 for more details) is concentrated in the Northern Plains and the Corn Belt regions (table 1.7). Together, these two regions contained half of all cropland used for crops in the contiguous 48 States in 2007.

Cropland used for crops declined over 1982-2007 by nearly 48 million acres, with every region experiencing declines between 2.0 and 9.2 million acres. The Northeast has experienced a long-term decline in cropland used for crops due to urban pressures and a comparative disadvantage in many crops. The Southeast region, however, experienced a larger rate of decline (36 percent over 1982-2007) than any other region.

The large decrease in cropland used for crops between 1982 and 1992 reflects an increase in land enrolled in Federal acreage reduction programs and the Conservation Reserve Program (CRP). These increases were authorized by the 1981, 1985 and 1990 Farm Acts. CRP was authorized by the 1985 act, began enrolling acreage in 1986, and accounted for a majority of diverted acres by 1990 as the Acreage Reduction Program was phased out (Nickerson et al, 2011; Hellerstein, 2006).

Table 1.7. Cropland used for crops, by region, 1982-2007[1]

Region	1982	1987	1997	2002	2007	Change 1982-2007
			Million acres			
Northern Plains	93.7	84.5	89.0	86.0	84.5	-9.2
Corn Belt	86.5	80.7	83.4	83.0	82.7	-3.8
Southern Plains	35.8	31.6	32.4	30.7	30.7	-5.1
Lake States	39.8	34.7	35.0	36.4	36.1	-3.7
Mountain	37.4	33.0	33.7	31.4	30.7	-6.6
Pacific	22.2	21.5	19.4	18.5	17.4	-4.8
Appalachian	19.3	16.6	16.9	17.1	17.3	-2.0
Northeast	13.6	11.1	10.7	11.4	11.0	-2.6
Southeast	14.6	10.4	11.6	9.8	9.4	-5.2
Delta States	19.2	16.5	16.3	15.7	15.2	-4.0
48 States[1]	382.6	337.4	348.6	339.8	334.9	-47.7

[1]Distribution may not add to totals due to rounding. Source: See citations in Nickerson et al. 2011.

Changes in the Mix of Major Crops

While cropland use in the United States has varied over the past three decades, the 8 most common crops—corn, soybeans, hay, wheat, cotton, sorghum, barley, and rice—have consistently accounted for the bulk of harvested acres. In 1982, farmers harvested over 316 million acres from these crops. By 1992—largely due to the increased enrollment of farmland in Federal acreage reduction programs and the CRP—only 276 million acres of these crops remained. As grain prices spiked in 1997, harvested acreage of these 8 crops rose to 298 million acres.

By 2007, the expansion of minor crops, and the increased planting flexibility introduced by the 1996 Farm Act allowing producers to fallow land that had formerly been maintained in more permanent cultivation, caused the level of harvested acres of major crops to drop to roughly 286 million acres (fig. 1.7). However, higher corn prices—from a strong demand for ethanol—resulted in a sharp increase in corn acreage in 2007 (and later years). Many farmers increased their corn acreage by changing their rotation between corn and soybeans, which lowered the harvested acreage of soybeans in 2007, but soybean acreage rebounded by 2008. The simultaneous net expansion of both corn and soybean acreage over the 2000-2009 period resulted from a complex set of land use changes, largely shifts from cotton, wheat and uncultivated hay acreage and an increase in double cropping (Wallander et al., 2011).

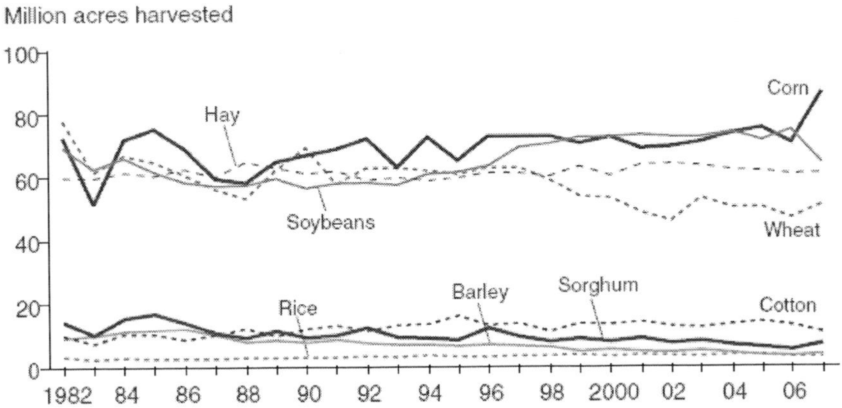

Source: USDA, Economic Research Service, compiled from USDA's National Agricultural Statistics Service, (Quick Stats, 1982-2007).

Figure 1.7. Acres of major crops harvested, 1982-2007 Million acres harvested.

Grazing Land

Land available for grazing includes grassland pasture and range (614 million acres in 2007), grazed forest land (129 million acres), and cropland used for pasture (36 million acres). Over 1982-2007, permanent grassland pasture and range made up about half of all agricultural land, with more than 90 percent of this land in the western half of the United States.

Grazed forest land includes acreage in open forest, land reverting to forest, and other forested areas with grass or other forage growth that supports at least some grazing. This type of grazing is common throughout the West, Southwest, and the South, and regional acreage variations are a function of the productivity of forested grazing, the demand for grazing, the amount of forest land, and factors such as species composition and stand density. In part, declines in grazed forest land since 1982 (table 1.5) are due to increases in forest stand density (restricting grazing possibilities) and improvements in both livestock feeding and forest management practices. These factors have been especially important in the South, where woodland grazing acreages have historically been high.

Cropland pasture, the smallest but generally most productive component of grazing acreage, comprises a relatively high percentage of total grazing land in the eastern half of the United States. While rising commodity prices may induce shifts from grazing to cropping uses, farmers tend to employ idled cropland first since it is generally more suited to crop production. Cropland pasture declined from 65 million acres in 2002 to 36 million acres in 2007.[3]

More Agricultural Land Is Subject to Urban Influence

Characteristics that make land attractive for farming (e.g., relatively flat land) also make it attractive for residential, commercial, and industrial development. While nationally only a small proportion of agricultural land has converted to urban uses, demand for urban uses is growing and, in contrast to other shifts in land use, land converted to an urban use rarely transitions back for agricultural purposes. During 1982-2007, NRI data indicate that most of the rural land converted to urban uses—over 11 million acres of cropland (roughly 3 percent of the cropland that existed in 1982), 12 million acres of pasture and rangeland (roughly 2 percent of the pasture/rangeland that existed in 1982), and 16 million acres of forest-use land—was previously in an agricultural use.

Agricultural land near residential concentrations is often in high demand for other uses, and this can change its relative profitability. As areas become desirable for nonfarm uses, characteristics unrelated to agricultural production (natural amenities, for example) become important determinants of farmland value (Barnard et al., 2003).

Proximity to urban areas can also alter farm operators' and landowners' resource allocation decisions. Rising land values can induce farmers to switch to higher value crops that increase returns to farming. More frequent contact with nonfarm neighbors can also increase production costs (e.g., increasing travel costs to farm fields due to traffic congestion or dealing with complaints about farming dust and odors), which can induce farmers to adopt new production methods. Farmers may also begin producing commodities that can be marketed through farmers' markets and "pick-your-own" operations, which shift transportation costs to buyers.

Nationally, in 2007, about 20 percent of the land in farms was subject to the effects of urban interactions. ERS estimates that an additional 4 percent of U.S. agricultural land became newly subject to urban influence between 1982 and 2000. Urban influence expanded fastest in Appalachia and the Southeast, where an estimated 13 percent of land became newly subject to urban influence.[4]

The Future?

The amounts of both farmland and farm labor employed in the United States have been declining for decades. Between 1982 and 2007, the decline in total farmland was about 5 percent, but the decline in cropland was 13 percent.

The aggregate supply of land—for farming and other uses—is ultimately fixed. If expected future demands begin to strain U.S. agriculture, farm operators can shift land out of other agricultural uses and into cropland. Nevertheless, there is a limit to how much range, pasture, and forest land would be productive as cropland.

The number of farms has stabilized in recent years, but only with growth in the number of very small farms whose operators depend on off-farm work and income. The amount of farm labor provided by hired laborers and operators declined by 30 and 40 percent, respectively, between 1982 and 2007, but agricultural production continued to grow due to heightened productivity.

Increased commodity prices—from, for example, world population growth or energy demands—could provide incentives to shift more land into crops and entice more people to enter farming. Nevertheless, future growth in farm production is likely to arise from continued productivity growth if historical trends continue.

Business Organization and Arrangements — and Risk

Farmers make many choices concerning their farm organization, and these choices have legal and economic implications. Organizational and marketing choices can also help reduce risk. For example, incorporating limits farm operators' legal liability. Larger farms (measured in terms of gross farm sales) tend to be complex—with extensive assets at risk—and operators of these farms incorporate more often than smaller farm operators, who may not find it worth the time or expense to use anything but the simplest form of organization available, the sole proprietorship.

Farming requires the control of land and other assets (livestock and machinery, for example), but these assets can often be rented—or hired on a custom basis—rather than owned, to avoid debt and the financial risks associated with the commitment of capital. Contracts can link farms to other businesses to market production and limit price or production risks. Often, farmers also have business arrangements with the Federal Government via crop insurance and other support programs to safeguard against production- or demand-driven risks to income.[1]

Business Organization

Farmers may use any of three traditional forms of business organization for legal purposes: proprietorship, partnership, or corporation. The simplest form of legal organization, the sole proprietorship, does not require any legal action to form, and the law makes no legal distinction between the proprietor (the owner of the business) and the business. The proprietor (the farmer or farm couple) controls the farm assets, is legally liable for the farm's debt and management decisions, and receives the income from the business. The proprietor reports the farm net income on his or her personal tax form (Johnson and Perry, 2001).[2]

More complex forms of legal organization—partnerships and corporations— allow multiple owners to work together. A single individual or family may not have the resources—management, labor, and assets—to run a commercially oriented farm. Partnerships and corporations allow people (not necessarily related to each other) to pool resources.

At least two people own a farm partnership and follow a legal agreement, which specifies the rules for managerial decisionmaking, the division of profits/losses, dispute resolution, and each partner's contribution of labor and capital. Each partner remains liable for the farm's debts and management decisions and reports his or her shares of the farm's net income on his or her personal income tax forms.

In contrast, a corporation is chartered under State law. It is an entity legally separate and distinct from its owners (the shareholders), who have no legal responsibility for the debts and decisions of the firm. However, officers of the corporation, who may also be owners, can be held liable for their actions as officers. Unlike a sole proprietorship or partnership, a corporation has a life of its own that does not end with a change in ownership or with the death of the owners.

Trends in Business Organization

Sole proprietorships, partnerships, or family corporations—where related people hold a majority of the stock—have consistently made up all but 1 or 2 percent of farms since 1982 (table 2.1). The distribution of farms among these categories has been stable, with sole proprietorships making up 86 to 88 percent of farms each year.

The same three forms of organization also accounted for more than 92 percent of agricultural sales over 1982-2007 (table 2.2), though with marked shifts in the distribution of sales. Family corporations and partnerships each increased their share of sales by 5 percentage points, while proprietorships' share shrank by 10 percentage points. Still, sole proprietorships accounted for about 50 percent of sales in 2007.

Nonfamily corporations—where related people *do not* own a majority of the stock—make up a relatively minor and stable share of farm numbers and sales. Nonfamily corporations, part of the "other organization" category in tables 2.1 and 2.2, accounted for 0.3 to 0.5 percent of all farms and 6 or 7 percent of agricultural sales each census year. Most nonfamily corporations are not large, publicly held companies. In 2007, 69 percent of nonfamily corporate farms had less than $250,000 in sales.

Table 2.1. Farms by business organization, selected censuses, 1982-2007

Business organization	1982	1987	1997	2007
		Percent		
Sole proprietorship	87.8	86.7	86.0	86.5
Partnership	10.0	9.6	8.9	7.9
Family corporation[1]	2.3	2.9	4.0	3.9
Other[2]	0.9	0.9	1.2	1.7
Total	100	100	100	100

[1]More than 50 percent of the stock is owned by persons related by blood or marriage.
[2]Nonfamily farms, estates or trusts, prison farms, grazing associations, American Indian Reservations, etc.
Source: USDA, Economic Research compiled from census of agriculture data.

Table 2.2. Farms product sales by business organization, selected censuses, 1982-2007

Business organization	1982	1987	1997	2007
		Percent		
Sole proprietorship	59.1	56.3	52.2	49.6
Partnership	16.3	17.1	18.1	20.9
Family corporation[1]	17.4	19.5	23.3	22.1
Other[2]	7.3	7.1	6.5	7.3
Total	100	100	100	100

[1]More than 50 percent of the stock is owned by persons related by blood or marriage.
[2]Nonfamily farms, estates or trusts, prison farms, grazing associations, American Indian Reservations, etc.
Source: USDA, Economic Research compiled from census of agriculture data.

Current Organization and Farm Size

While most operators organize their farms as sole proprietorships in the United States, the share of farms organized as partnerships and corporations steadily increases as farm size increases. Ninety-seven percent of the smallest farms (sales less than $10,000) are proprietorships, a larger share than any other sales class (fig. 2.1). Most of these very small farms actually are rural residences or retirement farms whose operator households receive much, if not all, of their income from off-farm sources. About 78 percent of the households in this sales class incur losses from farming, yet median household income for the sales class was $50,900 in 2007, or $3,600 more than the median for all U.S. households ($47,300). Given the scale of these farms and the small returns they provide, it may not be worth the expense or time to use any organization other than the proprietorship. By the time sales reach $1 million, however, only 51 percent of farms are organized as sole proprietorships.

Family Farms

USDA's Economic Research Service currently defines family farms as operations where the operator and individuals related to the operator—including relatives who do not live in the operator's household—own the majority of the farm business. Based on USDA surveys, the family-farm share of the farm count has remained fairly stable over the years, consistently falling between 97 and 99 percent of all farms from 1988 to 2007.[3]

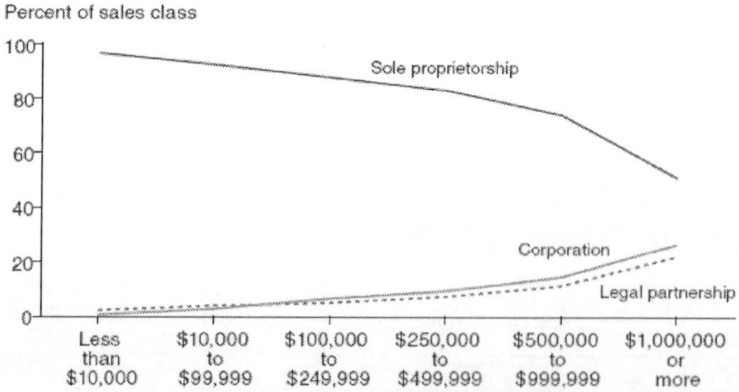

Notes: Sole proprietorships in ARMS include informal partnerships, such as those between spouses. The graph excludes the small "other" category, which includes estates, trusts, cooperatives, and grazing associations.

Source: USDA, Economic Research Service and National Agricultural Statistics Service, 2007 Agricultural Resource Survey, Phase III.

Figure 2.1. Business organization of farms by sales class, 2007.

Farming in the United States, therefore, is largely a family business, regardless of sales class. About 98 percent of U.S. farms were family organizations in 2007, and these family farms accounted for 82 percent of production (table 2.3). Family farms' share of the farm count and production drops as farm size increases. Yet, family farms still accounted for 86 percent of million-dollar farms and 69 percent of their production in 2007.

Table 2.3. Family farms by sales class, 2007

	Family farms' share of...	
Business organization	Farms	Value of production
		Percent
Less than $10,000	98.4	94.5
$10,000-$99,999	97.1	97.2
$100,000-$249,999	96.9	96.7
$250,000-$499,999	96.6	96.5
$500,000-$999,999	96.2	96.0
$1,000,000 or more	85.6	69.2
All farms	97.6	82.1

Source: USDA, Economic Research Service and National Agricultural Statistics Service, 2007 Agricultural Resource Management Survey, Phase III.

Limited Liability Companies (LLCs)

LLCs are a relatively new form of business organization allowed under State law. They were first authorized by Wyoming in 1977, then by Florida in 1982 (Keatinge et al., 1992). Other States followed after 1990 due to favorable income tax rulings, changes in tax structure favoring pass-through taxation, and State legislatures trying to attract businesses. The LLC form of organization provides business owners with limited liability for debts and actions of the business, management flexibility, and pass-through taxation (U.S. Dept. of Treas., IRS, 2008). LLCs must choose to file taxes as an individual (i.e., a proprietorship), a partnership, or a corporation, which means LLCs can be sorted into the more traditional business organization categories. Overall, LLCs accounted for only 3 percent of U.S. farms and 16 percent of production in 2007 (table 2.4), but made up a larger share of million-dollar farms (15 percent) and their production (25 percent). As farm size (measured by sales) increases further still, the LLC share of farms becomes particularly high. Operators organized 30 percent of $5-million farms as LLCs.

Part of a Larger Firm

Less than 1 percent of U.S. farms are part of a larger firm, such as a company that processes farm products. However, about 3 percent of million-dollar farms are part of a larger firm, and these farms account for 10 percent of the output of that sales class. Forty-four percent of these million-dollar subsidiaries specialize in livestock and another 39 percent specialize in high-value crops (vegetables, fruits, tree nuts, and nursery/greenhouse products).

Table 2.4. Limited liability companies (LLCs) by sales class, 2007

Sales class	LLCs' share of...	
	Farms	Value of production
	Percent	
Less than $10,000	1.7	2.5
$10,000-$99,999	5.1	5.6
$100,000-$249,999	3.7	4.1
$250,000-$499,999	5.4	5.4
$500,000-$999,999	8.2	7.8
$1,000,000 or more	15.3	25.3
All farms	3.3	16.2

Source: USDA, Economic Research Service and National Agricultural Statistics Service, 2007 Agricultural Resource Management Survey, Phase III.

Farm Business Arrangements

U.S. farms use a variety of business arrangements that link them to other firms and individuals. For example, farms may use marketing or production contracts to sell or remove the commodities they produce. Agricultural production requires the control of land and other assets, and renting these assets enables this without tying up capital.

Marketing and Production Contracts[4]

For farmers, contracts offer two main advantages. First, farmers can use marketing or production contracts to ensure outlets for their commodities, especially in thin markets or in markets for perishable products (see box, "Types of Contracts," p. 24). Second, contracts can reduce production and price risks. Production risks arise from unpredictable events, such as drought, frost, hail, and insect infestations in the case of crops; or animal disease, feed supply shortages, and extreme temperatures in the case of livestock. Price risks arise from unanticipated changes in output or input prices.

Both marketing and production contracts can minimize output price risk by making contract prices or fees independent of market prices. Production contracts can eliminate most input price risk since contractors provide the inputs that comprise most operating expenses. Contracts may also shift some of the production risk to processors. For example, a poultry production contract where the contractor is responsible for veterinary services may reduce production risks for the producer. Contracting, however, may introduce new risks to the producer. For example, the farmer may be unable to produce commodities that meet the quality or quantity benchmarks specified by the contract, which can result in noncompliance penalties.

Types of Contracts

A *contract* is an agreement—written or oral—between a farm operator (contractee) and another person or firm (contractor) to produce a specific type, quantity, and quality of agricultural commodity. Farmers typically use two types of contracts, *marketing contracts* and *production contracts*, both of which specify how the grower is to be compensated.

Marketing contract. Ownership of the commodity remains with the farmer during production. The contract sets a price (or a pricing formula), product quantities and qualities, and a delivery schedule. Contractor involvement in production is minimal, and the farmer provides all the inputs. For crops, the contract is finalized before harvest. For livestock, the contract is finalized before the animals are ready to be marketed.

Production contract. The contractor usually owns the commodity during production and the farmer is paid a fee for services rendered. The contract specifies farmer and contractor responsibilities for inputs and practices. The contractor often provides specific inputs and services, production guidelines, and technical advice. In livestock contracts, for example, contractors typically provide feed, veterinary services, transportation, and young animals. The contract is finalized before production of the commodity.

Source: MacDonald and Banker (2005) and MacDonald and Korb (2008).

Contracts Since the 1990s[5]

The share of U.S. agricultural production under marketing or production contracts grew between 1991 and 2000, but stabilized at around 40 percent afterward (fig. 2.2). In contrast, the share of farms with contracts has been stable since 1991, with roughly 1 in 10 farms using either type of contract. About 8-10 percent of farms have marketing contracts in a given year, while only 2-3 percent have production contracts. Note that changes in the annual production or value of specific commodities may affect the share of the total value of production under marketing and production contracts. For example, the all-commodity share of production under contract will rise in years in which commodities with above-average contract shares account for a larger share of agricultural production.

Estimates of contract production also vary from year to year because the data are drawn from random samples of farms. The rest of the discussion of contracting focuses on specific commodities under contract—rather than aggregate production—which makes fluctuations from random sampling more pronounced. Thus, survey years are combined to smooth out these annual fluctuations (MacDonald and Korb, 2008).

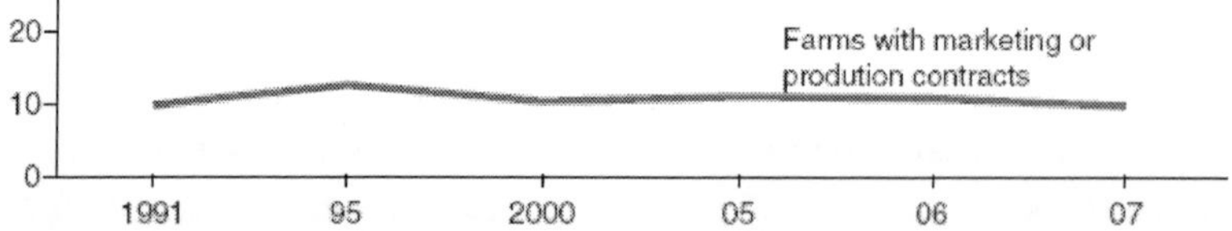

Notes: Data are not available for all years in the 1991 to 2007 period shown. Data include both marketing and production contracts.

Source: USDA, Economic Research Service and National Agricultural Statistics Service, 1991-1995 Farm Costs and Returns Survey and 1996-2007 Agricultural Resource Management Survey.

Figure 2.2. Use of contracts, 1991-2007.

Changes in Contract Production, by Commodity

Although the share of aggregate farm production under contract has stabilized at around 40 percent, the aggregate data mask large changes for specific commodities or commodity groups. In particular, the share of livestock production under contract grew from 33 percent in 1991-93 to 50 percent in 2006-07 (fig. 2.3), led by hogs and cattle under production contracts. Production under contract was more stable for dairy production (mostly marketing contracts) and poultry and egg production (generally production contracts). The share of crops under contract—typically marketing contracts—has remained relatively steady. Between 1991 and 2007, the share of crop production under contract has fluctuated between 25 and 30 percent (fig. 2.4). The shares of corn, soybeans, and wheat grown under contract have grown, although contract production of these crops remains below the overall crop average. We might expect more field crops—like corn, soybeans, and wheat—to shift to contracts if recent variability in crop prices continues, since both producers and processors might find production planning easier with more stable prices or pricing mechanisms.

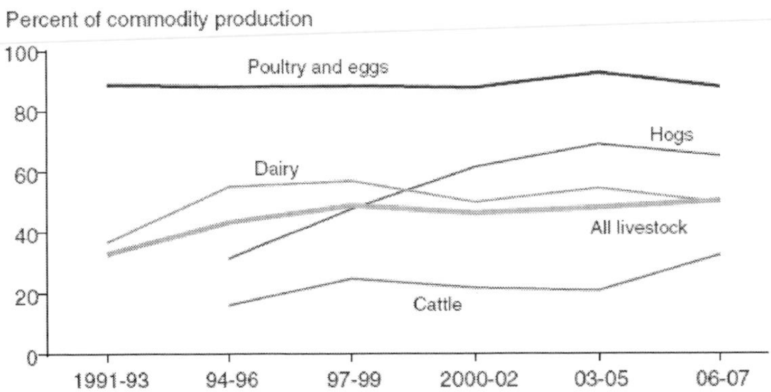

Note: Data include both marketing and production contracts.

Source: USDA, Economic Research Service and National Agricultural Statistics Service, 19911995 Farm Costs and Returns Survey and 19962007 Agricultural Resource Management Survey.

Figure 2.3. Production under marketing or production contracts for selected livestock species, 1991-2007.

For example, corn prices averaged $2-$3 per bushel from 2001 to late 2007, but have ranged from just over $2.00 to roughly $5.50 since. Similar volatility has occurred for wheat and soybeans (USDA, NASS, 2011).

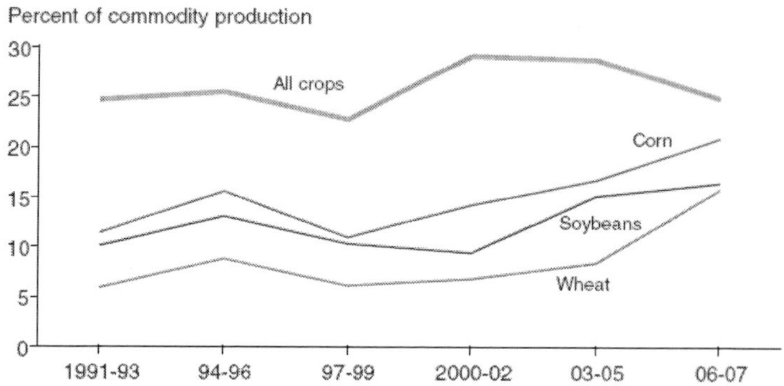

Note: Data include both marketing and production contracts.

Source: USDA, Economic Research Service and National Agricultural Statistics Service, 19911995 Farm Costs and Returns Survey and 19962007 Agricultural Resource Management Survey.

Figure 2.4. Production under marketing or production contracts for selected crops, 1991-2007.

Accessing Land

Land ownership is very common among U.S. farm operators: 84 to 96 percent of the farms in each sales class own some land (fig. 2.5). However, three-fourths of farms with sales of at least $100,000 report renting land. Renting farmland allows a farm operation to expand without incurring the risks associated with debt and the commitment of capital. Renting also enables rapid response to changing commodity markets. Farms with sales above $100,000 also report the most frequent use of share rental arrangements, ranging from 20 to 30 percent of the farms in each sales class. This reflects heavy concentrations of grain producers (who use share rental agreements extensively) in these sales classes. Forty-two percent of all farms with sales above $100,000 specialize in cash grains, accounting for 65 percent of all U.S. farmland rented for shares.

Nonoperator Landlords

Most of the farmland rented by farmers is leased from people who are not currently farm operators themselves. In the 2007 ARMS, 277,500 farm operators reported renting 62 million acres of farmland to others, which accounted for only 16 percent of the 395 million acres of rented farmland. Nonoperator landlords provided the rest of the rented land. Information about nonoperator landlords is sparse. The Agricultural Economics and Land Ownership Survey (AELOS) of 1999 still provides the most current, detailed information about nonoperator landlords and their farmland (USDA, NASS, 2001). Ninety-five percent of nonoperator land-lords were individuals/families or partnerships in 1999. Of these unincorporated landlords, 55 percent were at least 65 years old and another 11 percent were between age 60 and 64. Nonoperator landlords altogether provided 51 percent of U.S. cropland—excluding orchards, vineyards, and pastured cropland—used by farmers in 1999.[6] However their share of other types of land is substantially less including 23 percent of orchards and vineyards, 31 of pastureland, and 38 percent of woodlands not pastured.

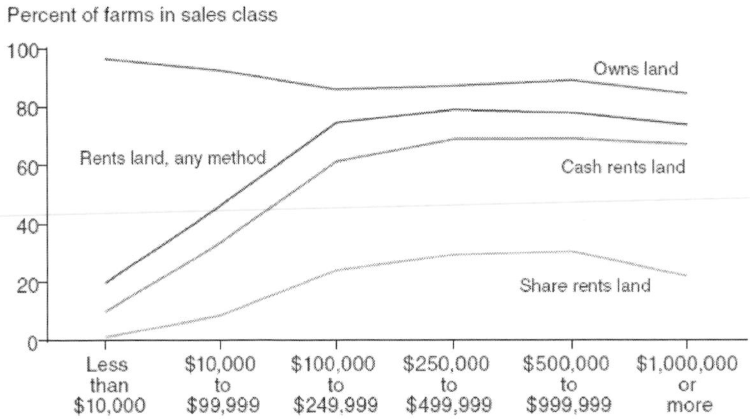

Notes: "Rents land, any method" includes land rented free of charge, not shown separately. Individual farms may rent for both cash and shares. Thus, the sum of the percent renting for cash and the percent renting for shares may total to more than the percent renting under any method. Similarly, farms may both own and rent land, so farms owning land plus farms renting land under any method sums to more than 100 percent.

Source: USDA, Economic Research Service and National Agricultural Statistics Service, 2007 Agricultural Resource Management Survey, Phase III.

Figure 2.5. Methods of accessing land by sales class, 2007.

Trends in Leased Farmland

Despite the prevalence of renting among farms with sales totaling at least $100,000, the number of acres leased by farmers actually declined by 13 percent (or 54 million acres) between 1992 and 2007—after stabilizing between 1982 and 1992 (fig. 2.6). Three-fourths of the decline in leased farmland came from tenants (operators who don't own any land), despite their smaller numbers relative to part owners. By 2007, tenants' share of leased farmland had declined to 23 percent from 30-32 percent in the 1980s and early 1990s, extending the long-term decline in tenant acres since 1935 (Janssen, 1993; USDA, NASS, *Agricultural Statistics*, 2008).

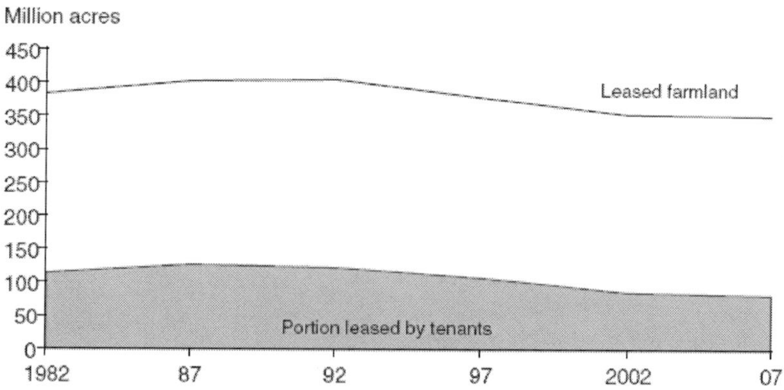

Source: USDA, Economic Research Service, compiled from census of agriculture data.

Figure 2.6. Leased farmland, 1982 to 2007.

One factor in the decline of rentals for both tenants and part-owners may be increasing rental costs as parcels of land available to rent become smaller over time because of division among heirs (Raup, 2003). Smaller parcels increase transaction costs for operators assembling land to expand their operations.

Accessing Other Assets

Farmers require more than just land to operate; access to other assets—such as machinery and livestock leases and custom work—can be vital to their operations. Although only 2 percent of the smallest farms (sales of less than $10,000) leased machinery in 2007, the share increases steadily with sales to 28 percent for million-dollar farms (fig. 2.7).

Custom work is any agricultural service farmers receive by paying for machinery and labor as a unit. Examples include fertilizer and pesticide application, sheep shearing, calf branding, land tillage, hay baling, and crop or livestock hauling (Aakre, 2005). Custom work is much more common than machinery leasing. More than half the farms with sales of $100,000 or more used custom work in 2007 (fig. 2.7).

Livestock may also be rented to avoid committing capital. For example, a farm operator with pasture but no cattle might choose to rent cows—rather than buy them—and pay the rent in cash or with a share of the calves produced (Doye et al., 2009). Or a farmer with a cow herd may lease a high-quality bull to avoid the cost of buying one. As another example, almond growers rent colonies of bees for pollination rather than establish their own apiaries.

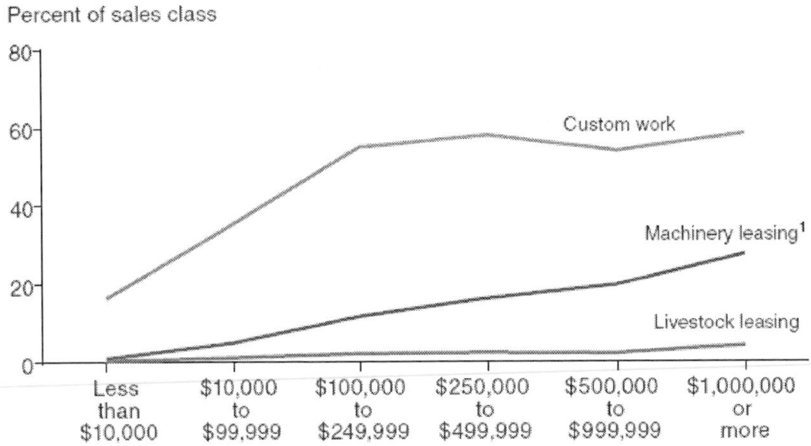

[1] Leasing of tractors, vehicles, farm machinery and equipment, and storage structures.
Source: USDA, Economic Research Service and National Agricultural Statistics Service, 2007 Agricultural Resource Survey, Phase III.

Figure 2.7. Selected methods to access resources, 2007 Percent of sales class.

Farms rent livestock infrequently, however, regardless of their sales levels. Only 1 percent of U.S. farms rented livestock, with little variation by farm size. Livestock rental is concentrated among high-value crop farms (43 percent of all livestock renters), which lease bee colonies to pollinate fruit, vegetable, or tree crops; and beef operations (34 percent).

Rental of machinery and the use of custom work among farms with sales of at least $100,000 (in 2007 dollars) has declined since the 1990s (fig. 2.8), though custom work was still performed on 56 percent of farms in 2007.

Federal Crop Insurance

While farmers can make individual decisions to mitigate the risk they face, such as using the legal organization to limit liability or engaging in contracts to reduce price risk, the Government also helps farmers by providing crop insurance and farm programs. Started in the 1930s as a response to the

Great Depression and the Dust Bowl, Federal crop insurance has only recently become a major tool that farmers use to mitigate the risk they face. Since few farmers enrolled in the program in the early years, lawmakers sought to make the program more attractive by passing legislation in the 1980 Farm Act that expanded crop insurance to cover more crops and regions of the country while also introducing subsidies to lower farmers' adoption costs. Although these incentives encouraged adoption, participation remained low, with farmers relying heavily on annual ad hoc disaster assistance bills to cover their crop losses.

[1] Sales are expressed in 2007 dollars, using the Producer Price Index for farm products to adjust for price changes.

[2] Leasing of tractors, vehicles, farm machinery and equipment, and storage structures.

Source: USDA, Economic Research Service and National Agricultural Statistics Service, 1991-1995 Farm Costs and Returns Survey and 1996-2007 Agricultural Resource Management Survey.

Figure 2.8. Selected methods used to access resources for farms with sales of $100,000 or more, 1991 to 2007.

In 1994, Congress passed the Federal Crop Insurance Reform Act (FCIRA). By both requiring crop insurance participation for Federal program eligibility and increasing subsidy levels to make crop insurance a more attractive option, participation greatly increased (fig 2.9).

In 1996, some of the program eligibility restrictions were removed, leading to an immediate decrease in coverage. However, crop insurance has expanded since then to introduce new insurance products while subsidies have also increased, encouraging further adoption.

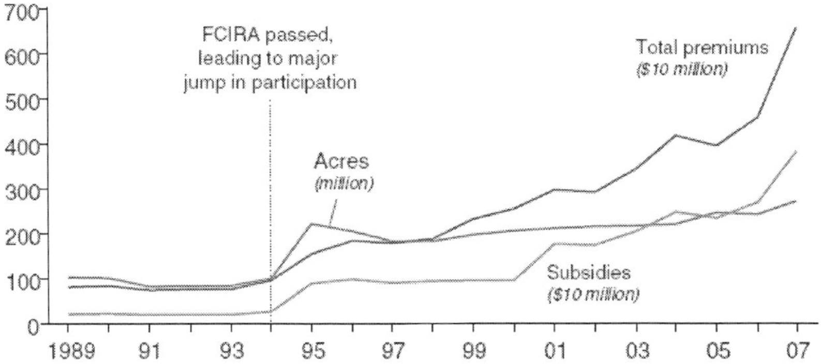

Source: USDA Risk Management Agency, Summary of Business Reports, 1989-2007.

Figure 2.9. Insurance coverage of all crops, 1989-2007.

To obtain crop insurance, a farmer must purchase a policy from 1 of 17 private-sector insurance companies. Although these companies sell and service the individual insurance policies, the Government develops and/ or approves the premium rates, administers the subsidies, and reinsures the commercial insurance providers (USDA, RMA, 2010).

Government Payments

Government policies have long been in place to address many farming issues, including land distribution, productivity, farmers' standard of living, marketing, and more recently, conservation. Government programs can have both positive and negative consequences. On the one hand, government intervention may help ameliorate market failures such as information asymmetries (which could, for example, cause insurance or credit markets to fail) or externalities (e.g., preventing undesirable agricultural outputs, such as manure, from polluting the Nation's air and water supplies). On the other hand, government intervention may cause farmers to modify their behavior—say, growing surplus crops or retiring land with marginal benefit to the environment—in ways that create inefficient resource allocations. Over time, policies have been adopted to support farm incomes, promote more environmentally beneficial production practices, account for increasingly high and fluctuating commodity prices, and address global pressures to reduce incentives that distort U.S. farmers' production decisions.

A Transition from Supply Management Programs toward the Market

Prior to the 1980s, Government policies typically fell into two categories: supply management and price support programs. Supply management programs attempted to control the supply of agricultural goods (often by restricting land use) which, in theory, allowed the Government to manipulate prices. Price supports tended to come in the form of production or marketing quotas and nonrecourse commodity loans, often creating price floors for specific commodities.

Beginning in the 1980s, supply management policies such as set-aside programs—whereby the Secretary of Agriculture made annual decisions regarding the percent of crop acreage that producers must idle for price support eligibility—fell out of favor, in part due to the difficulty of assessing their efficacy in increasingly global markets. As a result, the set-aside programs were phased out by the mid-1990s, though the Conservation Reserve Program still pays farmers to idle cropland for conservation purposes (Gardner, 2002).

Price support programs also underwent major changes beginning in the 1980s. Prior to this time, Congress supported various commodity prices with the creation of price floors by making loans to farmers after the harvest, based on predetermined loan rates. In turn, farmers gave their crops to the Government as collateral. If commodity prices were lower than the loan rate, the farmers kept the loan and forfeited their crops. If commodity prices were higher, the farmers bought back their crops and sold them at market prices.[7]

However, commodity surpluses proved costly to store and, by the mid-1980s, policymakers began to structure agricultural policies to more closely align support programs with the marketplace, creating incentives for farmers to market their commodities rather than forfeit their crops. By the 1996 Farm Act, Congress had replaced most of the older price support and supply control programs with new commodity programs aimed at achieving income support independent of farmers' production decisions (Dimitri et al., 2005).

Today's Programs

Two main types of programs categorize current Government payments to farmers: commodity-related and conservation. Commodity payments tend to reflect present or past production of specific commodities (mostly feed and food grains, cotton, and oilseeds). Commodity program payments make up the bulk of all Government payments to farmers. Between 1999 and 2008, commodity payments fluctuated between 74 and 93 percent of total farm program payments, although the share of commodity payments has dropped in more recent years (see box, "Farmers Rely Less on Commodity Payments in Recent Years," p. 32). The major commodity-related payments include the following:

- Direct payments (DPs) and countercyclical payments (CCPs) are based on the producer's historical production; DPs are based on fixed rates set in farm legislation, while current prices determine CCPs.
- Loan deficiency payments (LDPs) and marketing loan gains, collectively called "marketing loan benefits," and Milk Income Loss (MILC) payments are tied to current prices and/or production.
- Prior to the 2008 Act or disaster payments were made on an ad hoc basis to farmers when the Secretary of Agriculture determine that a region has qualified for disaster assistance. The 2008 ACT introduced a permanent emergency disaster program—the Supplemental Agricultural Disaster Assistance program.

Farmers Rely Less on Commodity Payments in Recent Years

While commodity program payments made up at least 70 percent of total government payments to farmers between 1996 and 2008, higher commodity prices and legislated decreases in payment acres for direct and Average Crop Revenue Election (ACRE) payments—combined with overall increases in conservation program expenditures in more recent years—have helped to reduce the share of commodity program payments to between 51 and 57 percent of all payments made to farmers. These market and policy changes have caused farmers to rely less heavily upon government payments as a source of both farm business and farm household income in 2009 and 2010 (with a similar forecast for 2011).

Conservation payments are designed to promote environmentally sound farm business practices, with programs geared toward reducing soil erosion, improving air and water quality, and maintaining and improving wildlife habitats.

- Land-retirement programs—the largest being the Conservation Reserve Program (CRP)—aim to retire environmentally sensitive land from production.[8]
- Working-land programs aim to enhance the farm operators' resource management on cropland and grazing lands currently in production.

In 2007, roughly 60 percent of all conservation payments consisted of land-retirement program funds dedicated to the CRP, although the majority of recent increases in conservation payments have accrued to the working-land programs.

Historic Trends

Although farm policy and farm programs have evolved since their introduction in 1933, farm programs have continually focused on supporting farm income and stabilizing commodity prices.

While their share of total gross cash farm income (GCFI)—the sum of receipts from the sale of crops, livestock, and farm-related goods and services, plus commodity (including disaster) and conservation payments—is low, these payments can be an important source of income for farmers participating in the programs.

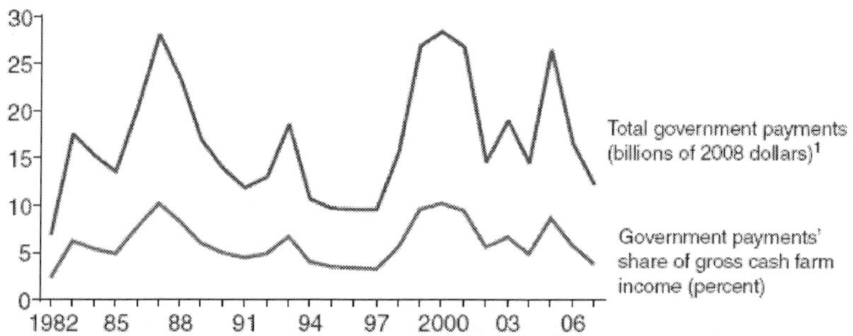

[1] Deflated with GDP chain-type price index. Deflating with the GDP price index shows the purchasing power of government payments relative to the rest of the economy Source: USDA, Economic Research Service, U.S. and State farm income data.

Figure 2.10. Government payments as a share of gross cash farm income, 1982-2007.

From 1982 to 2007, direct government payments' share of GCFI twice reached as high as 10 percent (fig. 2.10), first in 1987, just after the end of the farm financial crisis (generally dated from 1982 to 1986) and then in 2000, a result of large disaster and emergency payments in response to weak export demand and widespread crop failures (Gardner, 2002). Low commodity prices triggered high payments from 1998 to 2001, as well as in 2005, when government payments reached nearly 9 percent of GCFI. By 2007, government payments made up less than 4 percent of GCFI.

Annual Variation in Program Payments

Total payments made in any given year depend on programs' eligibility criteria, payment formulas, market and weather conditions, and the production and management choices of individual operators. As a result, they can fluctuate greatly from year to year (fig. 2.11). For example, in 2000—a poor year for farming—government payments totaled \$28 billion (measured in 2007 dollars), reflecting low commodity prices and widespread crop failures. In contrast, government payments were only \$12 billion in 2007, a relatively good year for farming. Some programs remain relatively static with foreseeable payment levels, while other programs fluctuate greatly depending on prices and the weather.

Variable Programs

Payments from programs designed to aid farmers with price and revenue risk can fluctuate greatly from one year to the next. The marketing assistance loan and LDP programs were introduced to encourage farmers to market their crops rather than forfeit them to the government. Similarly, countercyclical payments, introduced in the 2002 Farm Act, were designed to stabilize farm revenues. Periods of relatively low prices, such as those experienced in 2005, cause LDP, marketing loan gains, and countercyclical payments to rise, while high prices reduce these program payments.

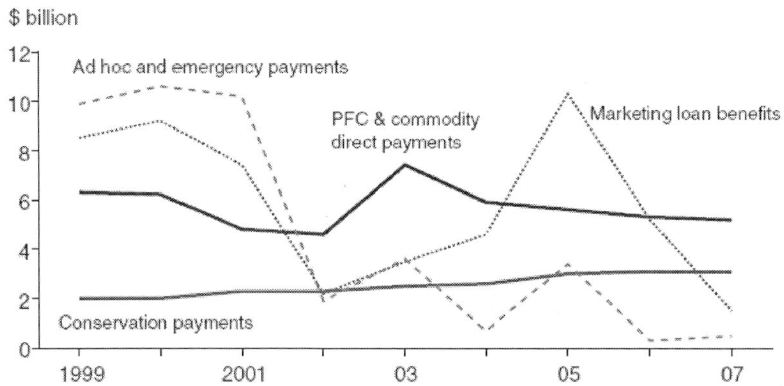

Note: Deflated with GDP chain-type price index. Deflating with the GDP price index shows the purchasing power of government payments relative to the rest of the economy.

PFC = Production flexibility contract payments.

Source: USDA, Economic Research Service, U.S. and State farm income data.

Figure 2.11. Annual variation in program payments by calendar year, 1999-2007.

The most variable commodity payments, the pre-2008 ad hoc and emergency payments, were triggered by severe weather or other natural (and occasionally, market) disasters. In most years, these payments have remained relatively low. When disaster strikes, however, ad hoc and emergency payments could easily eclipse other commodity payments. For example, in 1999 through 2001, annual disaster and emergency assistance payments totaled between $10 billion and $11 billion (in 2007 dollars), representing nearly 40 percent of all commodity payments in each of those years.

Stable Programs

In contrast to payments triggered by market swings or natural disasters, some commodity payments exhibit relative stability. Direct payments—introduced in 2002 and based on historical production and yields for individual farms—evolved from production flexibility contract (PFC) payments, introduced in 1996. Annual PFC and direct payments range within a fairly narrow band, from $5 billion to $8 billion (2007 dollars). These payments no longer fluctuate due to changes in prices, as did the commodity-specific payments they replaced (see box, "Transition to Production Flexibility and Direct Payments,").

As a rule, conservation program payments fluctuate much less than commodity program payments. Enrollment in the Conservation Reserve Program (CRP) requires a long-term commitment (currently 10-15 years), during which the landowner agrees to idle crop land and follow specified conservation practices such as planting environmentally friendly cover crops, creating grass waterways, or generating buffer strips. In turn, the landowner receives a predetermined "rental" fee from the Government based on the agricultural rental value of the land. The long-term nature of the CRP contracts lends stability to aggregate payments.

Before the introduction of the Conservation Reserve Program in 1986, conservation program payments made up only 2 percent of all government payments made to farmers. With CRP, total conservation payments rose to roughly 9 percent of government payments. Recent increases in funding for working-land programs such as the Environmental Quality Incentive Program (EQIP) and the Conservation Security Program (CSP) have helped raise total conservation program spending from $2.2 billion in 2003 to over $3 billion by 2007, more than a fourth of total agricultural program payments.

Transition to Production Flexibility and Direct Payments

Before 1996, most program payments were coupled to production decisions. These links were broken in the 1996 Farm Act, which tied PFC payments (and later, direct payments) to historical production and yields. After 1996, farmers have received roughly between $4 billion and $7 billion per year, irrespective of market conditions.

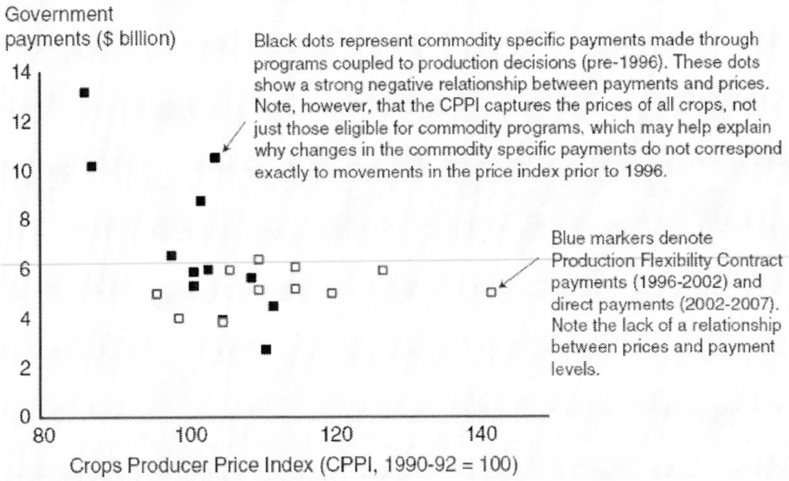

Source: USDA, Economic Research Service, Direct government payments, by program: 1996-2007.

Who Receives Government Payments?[9]

In 2007, total Government payments comprised almost 4 percent of total gross cash farm income (GCFI). This masks the importance of Government payments for some farms (fig. 2.12).

Because most support programs tie Federal assistance to specific commodities and production levels (either past or present), larger farms collect higher levels of Government payments. As farm size increases, however, GCFI tends to increase faster than Government payments, so the share of GCFI coming from Government payments tends to drop.

Overall, only 39 percent of farms participated in farm programs in 2007, ranging from 23 percent of farms with sales less than $10,000 to 79 percent of farms with sales of $250,000-$500,000. Participation rates increase with farm size largely because of their heavy participation in commodity programs.

Participation in conservation programs, in contrast, remains relatively low (14-29 percent in 2007) for all size groups.

According to the ARMS data, in 2007 operators of farms with sales above $250,000 harvested more than three-fourths of the total acres of program crops—barley, canola, corn, cotton, oats, peanuts, rice, sorghum, soybeans, and wheat. As a result, these farmers received roughly 71 percent of commodity-related payments. In contrast, farmers of operations with sales below $100,000 received 59 percent of all conservation payments, including 74 percent of land-retirement payments. Meanwhile, working-land conservation program payments followed the pattern of commodity program payments—farms with sales above $250,000 received nearly 60 percent.[10]

The Impact of Farm Commodity Program Payments

Understanding the impact of farm commodity program payments is no easy task. Many confounding factors are difficult to unravel and, as a result, research results are divided on how program payments affect farmers, their enterprises, and the subsequent impacts on farm household incomes. Even when consensus exists about *how* a payment affects farmers, estimates of the size of the effects often vary widely. Some payments may go straight into a farmer's pocketbook, directly increasing his or her household income. Other payments may take a much less direct route, and may, in fact, cause expenses to increase (such as land rental rates), decreasing the payment's overall impact on farmer household well-being.

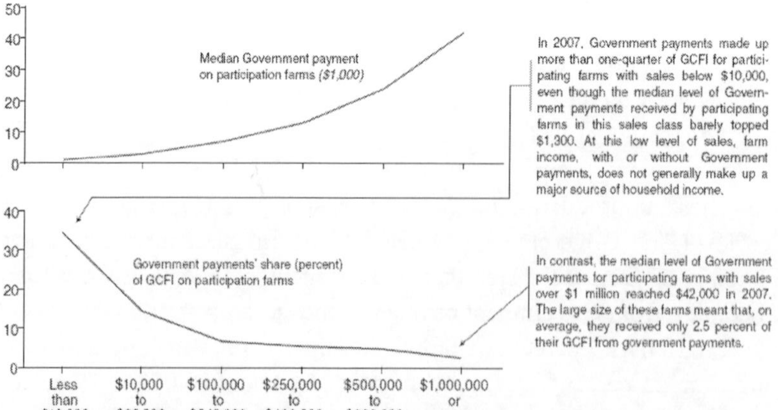

Figure 2.12. Government payments as a share of gross cash farm income by farm size, 2007.

Program payments have the potential to affect farmers, farmers' households, and their production decisions on many levels. For some farmers, payments may provide opportunities to increase the size of their operation. A steady stream of income may allow recipients to gain access to higher levels of credit or may allow them to increase their rental or purchase bids for land. This may provide opportunities for them to increase in size while driving out competition from smaller farms that don't have access to the same levels of capital, which can impact the overall structure of agriculture.

Owners of land—not necessarily farm operators—may be able to capture some, most, or even all of certain types of payments by raising land prices and rental rates.

As a result, certain policies aimed at increasing farm household income may not reach the intended target population, going to landlords instead.[11]

SPECIALIZATION AND CONCENTRATION OF PRODUCTION

A fundamental decision by farmers—what to produce—has wide-ranging implications for what the farm looks like, how it is organized, and its longterm success. Factors like agronomic concerns, market availability, the planning time horizon, and land quality influence this decision.

If overall cropping is too specialized, a farmer may lose everything in a particular year due to a specific disease or infestation. Producing a small number of commodities may also require a more intensive use of inputs; for example, continuous corn requires higher levels of fertilizers than a corn-soybean rotation. However, specialization also has its benefits. Farmers can become expert in growing just a few crops, maximizing output, minimizing the different types of capital (machinery, buildings, etc.) required for production, and economizing on time. As a result, farmers may reduce costs and capture some efficiencies by concentrating on the production of a smaller number of commodities or a specific phase in the production of a single commodity. Farmers must weigh the pros and cons of specialization and choose a level of specialization where they are comfortable. As a result, the level of specialization differs across producers.

Trends in Farm Specialization

While the Nation produces a wide variety of commodities, farms have become more specialized over the past 30 years—the share of farms producing a particular commodity has generally decreased (fig. 3.1). Despite these trends, the number of acres cultivated of particular crops has remained relatively constant. To examine these trends among farms with significant production, we focus on farms with sales above $1,000, eliminating a considerable number of farms: 254,000 in 1982 and 688,800 in 2007.[1]

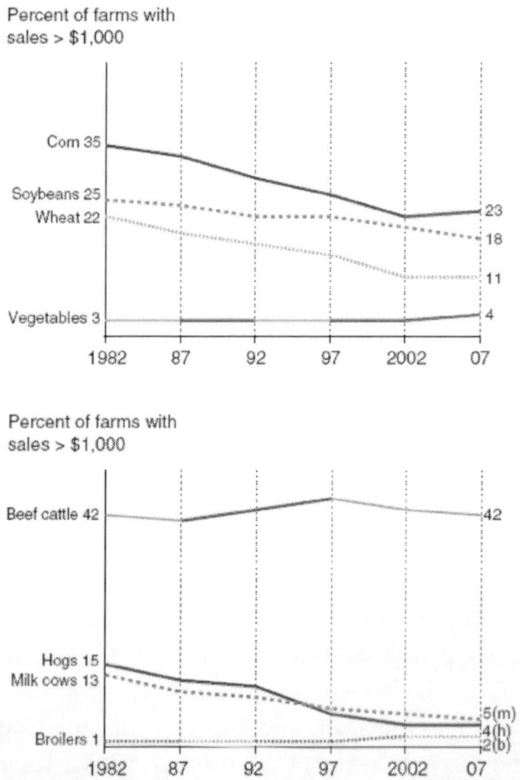

Note: the blue lines represent an upward movement in the trend while the black lines represent a drop in the trend line.

Source: USDA, Economic Research Service, compiled from U.S. Bureau of the Census, 1982 Census of Agriculture and USDA, National Agricultural Statistics Service, 2007 Census of Agriculture.

Figure 3.1. Recent trends in specialization among farms with at least $1,000 in sales, 1982-2007.

Of all farms that sold at least $1,000 worth of agricultural goods in 1982, roughly 35 percent grew corn; by 2007, fewer than 23 percent of these farms grew corn. The share of farms growing soybeans fell from more than 25 percent in 1982 to 18 percent in 2007, and the share of farms cultivating wheat dropped at rates similar to the share growing corn.

Livestock farms experienced similar trends over 1982-2007. While the share of farms raising broilers exhibited a small increase over the 25-year period, likely due to the significant increase in demand for the commodity, the percent of farms with hogs and milk cows fell considerably. The share of farms producing beef cattle, by far the most common commodity raised on farms, fluctuated around 42 percent throughout the period.

Specialization and Farm Size

While the share of farms growing a particular commodity has generally dropped over the past 30 years, specialization also appears to differ across farm sizes. Data from the Agricultural Resource Management Survey (ARMS) suggest that specialization at a given point in time decreases with farm size (Hoppe et al., 2007). In 2004, about three-fifths of family farms with sales of $100,000 or more produced 3 or more commodities, while 75 to 85 percent of the smaller family farms produced 1 or 2 commodities, possibly a function of the acres at their disposal.[2]

The relationship between farm size and specialization varies by commodity, particularly when viewed over time. Melhim et al. (2009b) found that between 1992 and 2002, the industry average specialization levels (or share of farms growing) dropped for apple, dairy, and wheat farms. However, the largest farms in these industries diversified the least over time (making them relatively more specialized in comparison to smaller apple and dairy farms). In contrast, for almost all farm sizes (except the smallest corn farms), beef and corn farms became more specialized over this timeframe. Despite the differences across commodities, the smallest farms generally remained the most specialized over the period examined (Melhim et al., 2009a/b).

Geographic Specialization

Specialization occurs not only at the farm level, but also at a State or regional level. Some geographic specializations are longstanding and were established by the late 1800s, reflecting local comparative advantages in the production of specific commodities (Cochrane, 1993). However, these geographic specializations can change over time, shifting production from one region of the country to another due to changes in technology (for example, pesticides that contributed to the elimination of the boll weevil), climate, markets (e.g., population growth in the West), and/or business arrangements.

If production becomes geographically concentrated, the byproducts of production, such as manure or chemical use (including nutrients and pesticides), also become more localized. These changes have implications both in terms of the Nation's waterways (the potential for water pollution) and for those living nearby (the potential for air pollution). These issues warrant a clear understanding of current geographic trends in commodity specialization.

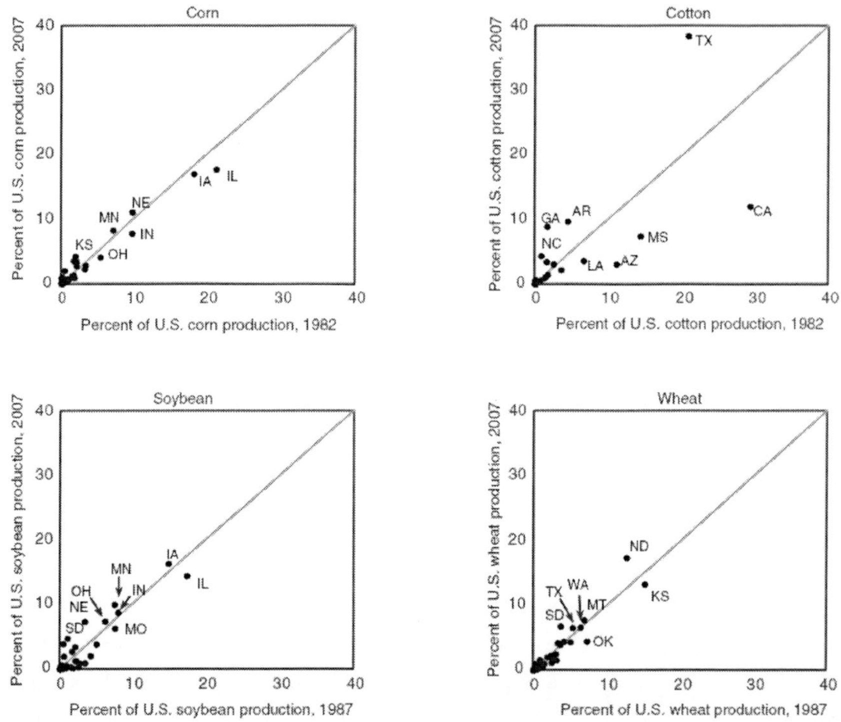

Each dot represents a State's share of production of a crop in both 1982 (horizontal axis) and 2007 (vertical axis). States with an increasing share between 1982 and 2007 lie above the 45 degree line.

Source: USDA, Economic Research Service, compiled from U.S. Bureau of the Census, 1982 Census of Agriculture and USDA, National Agricultural Statistics Service, 2007 Census of Agriculture.

Figure 3.2a. National geographic changes in selected crops, 1982-2007.

State Shares

Figures 3.2a and 3.2b (for crops and livestock respectively) show how U.S. production of various commodities has changed over time. Each dot represents a State's share of production of the commodity in both 1982 (horizontal axis) and 2007 (vertical axis). States with an increasing share of the commodity's production between 1982 and 2007 lie above the 45-degree line, while those with a decreasing share lie below it.

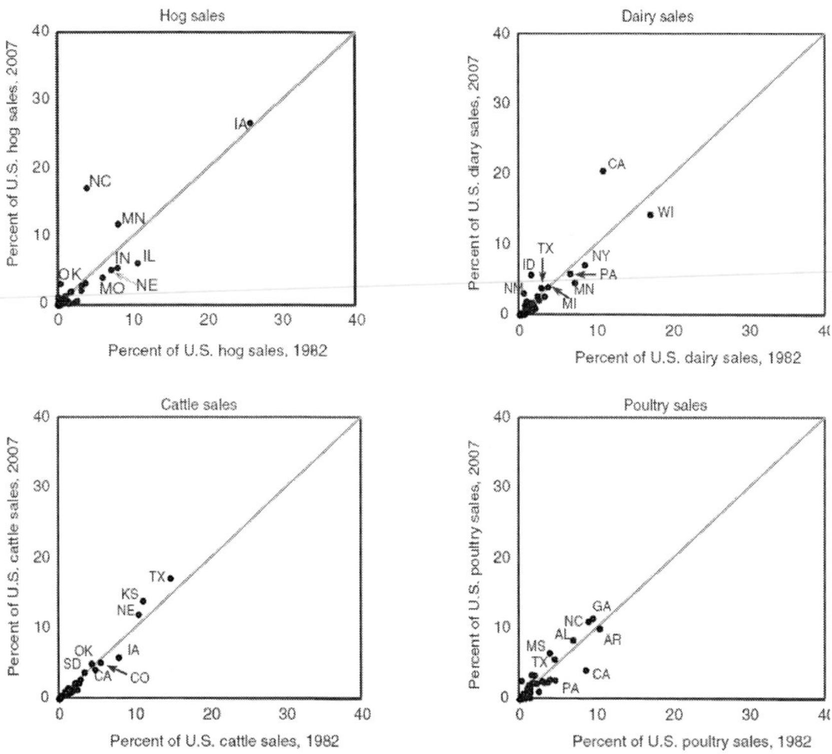

Each dot represents a State's share of production of a crop in both 1982 (horizontal axis) and 2007 (vertical axis). States with an increasing share between 1982 and 2007 lie above the 45 degree line.

Source: USDA, Economic Research Service, compiled from U.S. Bureau of the Census, 1982 Census of Agriculture and USDA, National Agricultural Statistics Service, 2007 Census of Agriculture.

Figure 3.2b. National geographic changes in selected livestock, 1982-2007.

Overall, changing technology, business arrangements, and climate have altered where commodities are produced. Corn and soybean production has shifted westward, cotton production has shifted eastward, and wheat production has shifted northward (fig. 3.2a). In livestock, hog production has shifted eastward, dairy production has shifted westward, cattle sales have moved toward the South and Central Plains, and poultry has moved southward (fig. 3.2b).

County Shares

Given the shifting of production around the country, examination of countylevel concentrations is necessary to unearth the potential for excessive manure production or chemical use. To do so, we calculate the fewest number of counties required to produce 25 percent, 50 percent, and 75 percent of total U.S. production of various commodities (see table 3.1).

Overall, substantial local geographic specialization occurred between 1982 and 2007 in the livestock sectors, including the cow-calf (cattle less than 500 pounds), dairy, fed cattle, and hog sectors. Broilers appear to be the exception, with larger numbers of counties required to capture 25, 50, and 75 percent of production.

Crops, on the other hand, have a much more mixed set of results. Rice and soybeans remain fairly stable over time, while wheat production appears to have concentrated geographically.

Meanwhile, larger numbers of counties produce 25, 50, and 75 percent of corn production as well as all levels of cotton production, likely due to higher plantings generally. Cotton's shift eastward—made possible by the eradication of the boll weevil in the South and drought in competing areas of the West—may also have played a role in its adding counties as production shifted from an area with larger counties (California) to the smaller counties of the Southeast.

Specialization has caused changes in how farm businesses are run and what the different farm sectors look like. At the individual farm level, the share of farms producing a particular commodity has dropped significantly, and farms do not produce as wide an array of outputs today as they did 100, or even 25, years ago. Production is also moving around the country, with shifts in many farm sectors toward production that is concentrated in fewer counties.

Table 3.1. Local geographic specialization: The minimum number of counties that produce 25, 50, and 75 percent of U.S. production, 1982 and 2007

			Production = number of head sold						*Production = number of milk cows*	
	Broilers		Cattle, < 500 lbs.		Cattle, fattened		Hogs		Milk cows	
	1982	2007	1982	2007	1982	2007	1982	2007	1982	2007
25%	15	26	186	77	22	15	71	21	50	17
50%	45	72	556	367	90	46	214	81	174	80
75%	112	156	1186	948	278	139	508	220	511	246
100%	3,034	2,046	3,058	3,039	3,022	2,539	3,031	2,918	3,034	2,529
					Production = acres harvested					
	Corn for grain			Cotton	Rice		Soybeans			Wheat
	1982	2007	1982	2007	1982	2007	1982	2007	1982	2007
25%	95	99	11	15	9	8	100	105	64	50
50%	244	257	36	47	21	17	265	268	178	139
75%	505	523	85	123	39	33	530	509	416	305
100%	2,825	2,634	615	627	157	135	2,182	2,039	2,803	2,481

This table shows the minimum number of counties that generate 25, 50, and 75 percent of production within the various agricultural sectors, along with the total number of counties that produce the commodity. Numbers in regular typeface represent industries that are consolidating – with fewer counties required to hold the requisite percent of production. Values in red typeface represent industries that are spreading out more – with larger numbers of counties required to hold the requisite percent of production. For example, in 1982, the 71 largest hog-producing counties generated 25 percent of the hogs sold in the United States. By 2007, production had consolidated, so now only the largest 21 hog-producing counties were required to capture 25 percent of the hogs sold in the United States. In contrast, in 1982, 15 counties held one-fourth of all broiler production. By 2007, this number had expanded to 26 counties (although the total number of counties with broilers did drop from 3,034 to 2,046).

Source: Economic Research Service calculations based on U.S. Census Bureau, 1982 Census of Agriculture; and USDA, National Agricultural Statistics Services, 2007 Census of Agriculture.

Trends in the Size Distribution of Farms and Sales

We use sales to measure economic activity rather than the other commonly used measure, land (Hoppe et al., 2010). Farmland varies in quality, can be farmed at different levels of intensity, and can produce a wide variety of products.

Thus, the value of production per acre can vary greatly from farm to farm. To properly examine the change in farm sizes over time, we need to adjust sales for price changes. Any change in sales is made up of two components: a price change and a quantity change.

To eliminate price changes and to compare sales levels accurately over time, we use the Producer Price Index for Farm Prices (PPIFP) to express every year's farm sales in 2007 dollars. This produces quantity, or real, changes in farm size.

Distribution of Farms and Sales

Farm numbers and acres per farm stabilized in recent decades largely because of an influx of farms with annual sales less than $1,000—point farms—and a smaller increase in the number of farms with sales of $250,000 or more, nearly offsetting a large decrease in the number of farms with sales from $10,000 to $249,999 (table 3.2).

Farms with sales of at least $1 million more than tripled between 1982 and 2007, though at 55,500 total they still represented just 3 percent of the U.S. farm count in 2007.

The distribution of farm sales also shifted upward to million-dollar farms (fig. 3.3), whose share of market value increased from 27 percent in 1982 to 59 percent in 2007. In contrast, farms generating agricultural products with a market value of less than $10,000 consistently accounted for only 1 or 2 percent of the U.S. total sales between 1982 and 2007, despite their growing numbers. Farms with sales of $10,000-$249,999 experienced the largest loss in market share, approximately 27 percentage points. Nevertheless—despite their dominance of the production landscape—too many million-dollar farms exist for any single farm to exert market power (Hoppe et al., 2008).

Production per Farm

The growing number of both large commercial farms and very small farms that produce little in the aggregate means that production is shifting from a large number of smaller farms to fewer, larger farms—measured by sales — as the number of midsized farms declines. The shift of production to fewer and larger farms has been underway for generations. For example, in 1900, it took almost 1 million farms in the United States to generate half of all the market value of agricultural products sold (Peterson and Brooks, 1993). By 1987, fewer than 76,000 farms were needed and by 2007, fewer than 33,000 farms produced half of the market value of agricultural goods sold (fig. 3.4).

Table 3.2. Number of farms, by constant-dollar class, 1982 and 2007

Sales class (2007 dollars)[1]	1982	2007	Change
	Number		*Percent*
Total farms	2,240,976	2,204,793	-1.6
Less than $10,000	954,349	1,319,161	38.2
Point farms[2]	254,097	688,834	171.1[3]
$1,000-$9,999	700,252	630,327	-10.0
$10,000 to $249,999	1,137,892	675,973	-40.6
$250,000 to $999,999	132,544	154,150	41.0
$1,000,000 or more	16,191	55,509	242.8

[1]Sales classes are defined in 2007 dollars, using the Producer Price index for farm products to adjust for price changes.

[2]Point farms have sales below $1,000 (current dollars), but are still considered farms because they would be expected to normally sell at least that much. Point farms are identified using current dollars because the minimal level of sales in the farm definition is not adjusted for price changes.

[3] See box, "Counting Farms: Methodological Changes"

Source: USDA, Economic Research Service, compiled from U.S. Census Bureau, 1982 Census of Agriculture and USDA, National Agricultural Statistics Service, 2007 Census of Agriculture.

However, farmers running these 33,000 farms have not notably increased the average number of acres they operate. The typical (average) size of these large farms has plateaued, likely reflecting a combination of production efficiencies and managerial constraints. The number of acres (on the farms generating half of the Nation's market value of agricultural products sold) required to generate these levels of sales has remained relatively stable at 2,500 to 3,000 acres per farm since 1987. While the average acreage of these farms has plateaued, their share of total land in farms dropped from more than 22 percent in 1987 to roughly 10 percent by 2007. Farms of this size often produce high-value crops (vegetables, fruits, tree nuts, and nursery/greenhouse products), beef (mostly feedlots), dairy products, hogs, and poultry and eggs—commodities that can be produced on relatively few acres (Hoppe et al., 2008).

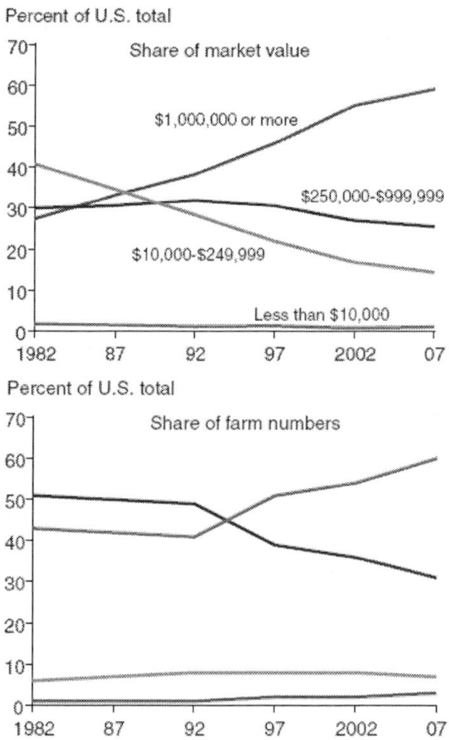

Figure 3.3. Share of market value of agricultural products sold and farm numbers, by constant-dollar sales class,[1] 1982-2007.

While the previous measure shows that the production per farm has, on average, increased greatly over time, we next explore whether or not these increases held across a wide spectrum of agricultural sectors by generating wo indices: an index of production and an index of the number of producers. By comparing the two indices, we can explore how levels of production changed with changes in the number of farms producing the commodity. The divergence of the two indices in figure 3.5 shows that the large increases in production per farm held across many different commodities over time.

For example, while the number of broiler producers dropped marginally from 1982 levels, broiler production increased by over 250 percent from 1982 to 2007. While the production per farm has generally increased across a wide variety of outputs, important differences exist between commodities. Fed-cattle production, unlike broilers, remained relatively stable while the number of producers dropped precipitously from 1982 levels. In other commodities—such as dairy and tobacco—both production and the number of producers decreased from 1982 levels. Nevertheless, across all commodities, with few exceptions, production per farm increased, often substantially.

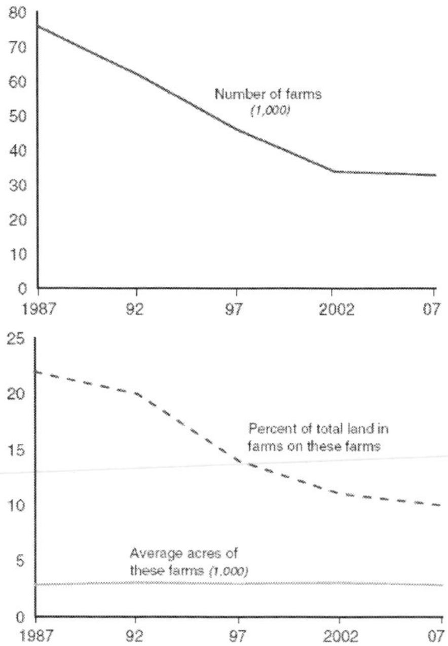

[1]This figure shows the smallest number of farms that account for 50 percent of farm product sales, measured as the market value of agricultural products sold, excluding government payments.

Source: USDA, Economic Research Service, compiled from Census of Agriculture data.

Figure 3.4. Number of farms that account for half of the year's farm product sales1, their average acreage, and their percent of total land in farms.

Increasing Size of Enterprise

While we have evidence that production per farm has increased over time, we need a measure to show how production is shifting to ever larger farms. Because production is heavily skewed to large commercial farms, comparisons of the mean (or average) farm size are at best, uninformative, and can easily be misleading. One way to explore changes in farm size over time is to use physical measures of outputs—such as bushels, head of animals sold, or herd size—eliminating the need to control for price changes. One measure

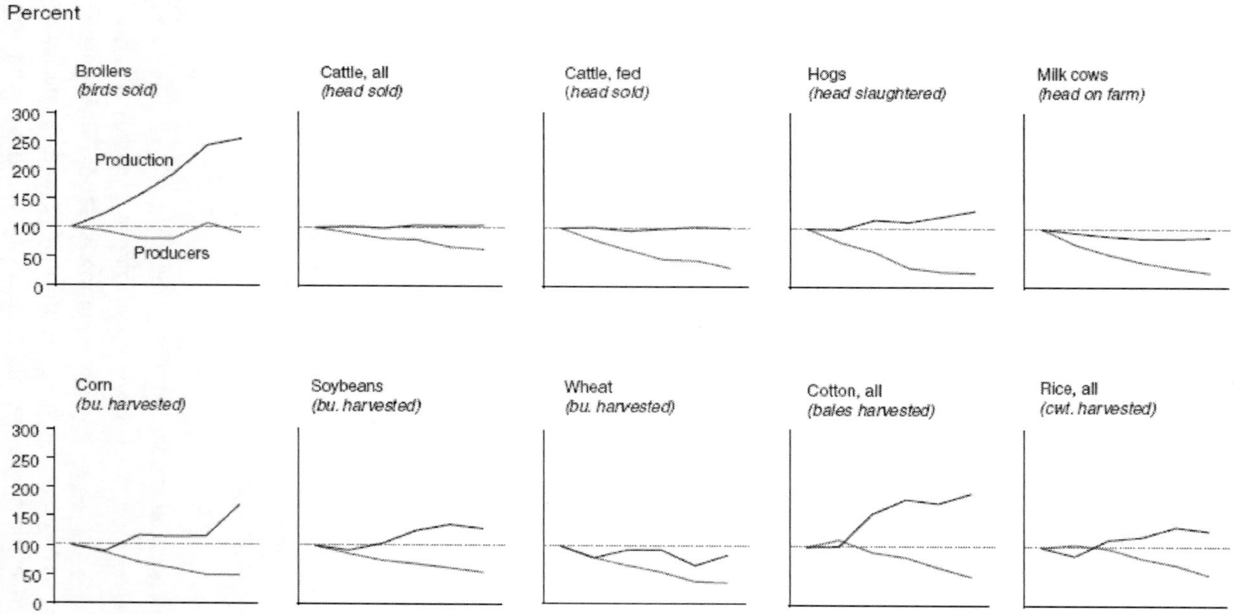

Figure 3.5. (Continued).

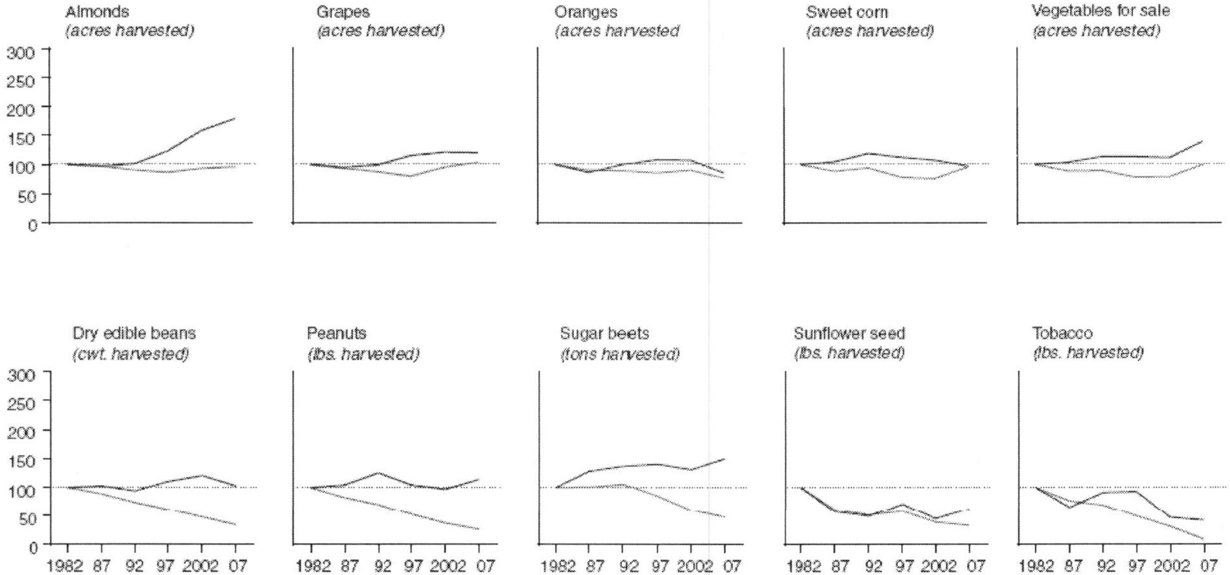

Note: Plots of the number of producers (blue lines) and total production levels (black dotted lines) relative to 1982 levels for various commodities. These plots show how, over time, production is getting concentrated into fewer, larger farms.

Source: USDA, Economic Research Service, compiled from U.S. Census bureau, 1982 Census of Agriculture and USDA, National Agricultural Statistics Service, 2007 Census of Agriculture.

Figure 3.5. Production index and number of producers index, 1982 = 100.

Table 3.3. Mid-aggregate enterprise sizes for selected commodities, 1987-2007

Selected commodity	Census year					Change	
	1987	1992	1997	2002	2007	1987 to 2007	
	Mid-aggregate annual sales1 (head per farm)					*Percent*	
Poultry/livestock:							
Broilers	300,000	384,000	480,000	520,000	681,600	127	In 1987, half of all hogs were produced on farms that sold 1,200 head of hogs or more. Significant changes in the hog sector led to half of all hogs being produced on farms selling 30,000 hogs or more by 2007. This was by far the largest increase experienced by any major agricultural commodity.
Hogs	1,200	1,880	11,000	23,400	30,000	2,400	
Fattened cattle	17,532	23,891	38,000	34,494	35,000	100	
Cattle, < 500 lbs.	50	60	65	84	128	156	
	Mid-aggregate herd size2 (head per farm)						
Dairy production	80	100	140	275	570	613	
	Mid-aggregate acres harvested3 (acres per farm)						
Field crops: Corn	200 300 350 450 600					200	
Soybeans	243	300	380	480	490	102	
Wheat	404	562	693	784	910	125	
Cotton	450	605	800	920	1,090	142	
Rice	295	400	494	607	700	137	Tomatoes provide a good example for why the mid-aggregate enterprise size makes sense, as opposed to using simple means. By 2007, the average tomato enterprise size dropped to 17 acres, down from 26 acres in 1987.
Vegetables:							
Asparagus	160	200	200	236	240	50	
Lettuce	949	1,168	1,461	2,225	1,815	91	
Bell peppers	88	130	180	200	300	241	
Potatoes	350	422	556	810	990	183	
Sweet corn	100	120	173	222	250	150	
Tomatoes	400	450	589	700	820	105	

Selected commodity	Census year					Change 1987 to 2007	
	1987	1992	1997	2002	2007	*Percent*	
Tree crops:							In contrast, because production had been shifting to larger farms over time, the midaggregate enterprise size had grown from 400 acres in 1987 to 820 acres by 2007, a 105-percent increase.
Apples	83	94	122	129	146	76	
Almonds	203	234	292	361	450	122	
Oranges	450	732	769	1,015	1,113	147	
Peaches	92	95	100	105	120	30	

Note: Census records do not have all the data necessary to derive the mid-aggregate enterprise size prior to 1987.

[1] Mid-aggregate head sold. Half of the sales of a given species were from farms with more sales and half were from farms with sales less than the mid-aggregate total.

[2] Mid-aggregate head of dairy cows as of December 31 of the census year. Includes dry cows and cows in milk.

[3] Mid-aggregate acres harvested. Half of all harvested acres of a commodity were on farms harvesting more acres and half were on farms harvesting fewer acres than the mid-aggregate total.

Source: USDA, Economic Research Service, compiled from census of agriculture data.

expressed in physical units is enterprise size.3 We adopt a measure coined by Lund and Price (1998), the "mid-aggregate point," to calculate the enterprise size where half of production takes place on enterprises larger than the mid-aggregate point and half of production takes place on smaller enterprises (table 3.3).[4]

Livestock

The increasing size of livestock operations is well documented and is reflected in our estimates of the mid-aggregate enterprise size. For example, in 1987, the mid-aggregate broiler enterprise sold 300,000 birds. By 2007, mid-aggregate annual sales had risen 127 percent to 681,600 birds per farm.

The most dramatic growth, however, occurred in hog and dairy production. Mid-aggregate hog sales went from 1,200 head in 1987 to 30,000 head by 2007, a 2,400-percent increase, while dairy farms underwent a 613-percent increase.

Crops

Production has also shifted to larger farms for crop enterprises over the last 30 years. The mid-aggregate enterprise growing corn grew from 200 acres of corn per farm in 1987 to 600 acres by 2007, a 200-percent increase. Similarly, enterprises producing cotton grew from a mid-aggregate size of 450 acres of cotton to nearly 1,100 acres in 2007, an increase of more than 140 percent. Even fruit and vegetable enterprises experienced significant increases over time. Farms with bell peppers, for example, increased their mid-aggregate enterprise size by over 240 percent, from 88 acres in 1987 to 300 acres by 2007. Other fruit and vegetable enterprises—such as almond, potato, and tomato—also more than doubled their midpoint size (table 3.3).

What Are Some of the Drivers of Change?

One reason farmers specialize is to take advantage of efficiencies and lower costs. By specializing in fewer commodities, farmers can become more expert at production, current in technology, and adept at managerial skills.

Despite the trend toward larger farms and greater specialization, there are advantages to diversifying. Agronomic concerns may lead a farmer to produce multiple crops in rotation, such as corn-soybeans. Corn uses nitrogen (N) to grow, while soybeans fix N into the soil and ready the soil for the next corn crop. Also, diversifying can allow for the full use of fixed inputs. For example, labor, land, or machinery may be idle at certain points in the year for a single crop—planting crops with different lifecycles can minimize the idling of inputs. Additionally, producing multiple commodities can mitigate risk if the outputs are either negatively correlated with each other (if one commodity fares poorly, the other fares well) or independent of each other (which does not perfectly insure against loss, but reduces exposure to catastrophic losses). Recent research, however, suggests that diversification is not widely used as a risk management tool by farmers (O'Donoghue et al., 2009a).

Farmers can use contracts to produce commodities for specified fees, transferring risk to the contractor who also (in the case of production contracts) provides many of the production inputs. This enables farmers to save time and resources, allowing them to expand their production of the contract commodity or devote time to other commodities or the pursuit of offfarm income. Farmers may also contract with custom producers to free up resources. For example, some dairy farms acquire cropland for feed production, yet rely upon custom producers for all cropland tasks.

Economies of scale, technological advances, and government policies may also help explain why production per farm is increasing and why production is becoming more specialized over time. Economies of scale occur when average costs fall as farm size increases, strong evidence of which has been found in livestock operations. Large-scale crop operations, however, do not appear to benefit from such economies. Growth in crop enterprises appears to stem from substantial cost advantages from the intensive use of fixed resources (in particular, operator and family labor). For example, larger, faster equipment has allowed producers to manage much larger farms over time. In 1970, an operator could plant 40 acres of row crops per day, planting 4 rows at a time at 2 miles per hour (mph), and could harvest 4,000 bushels per day running a 4-row harvester for 12 hours per day. By 2005, a producer could plant 420 acres per day, planting 16 rows at 6 mph, and harvest 30,000 bushels per day, running a 12-row harvester. In 2012, it is expected that farmers will be able to plant 945 acres per day using a 36-row planter at 6 mph and harvest 50,000 bushels per day with a 16-row harvester (Bechdol et al., 2010).

To realize cost advantages, farmers must make substantial capital investments. Strong economic incentives exist for farms to grow larger and for production to continue to shift to them; evidence of these incentives can be seen by examining the operating profit margin of different sized farms.

Average operating profit margins peak for farms with at least $1 million in sales, giving these farms a competitive edge (table 3.4). Farms with sales below $100,000 earn negative operating profit margins, on average.

Increases in productivity due to new technology may also lead to consolidation. Often, due to cost averaging over more output, larger farms can more readily take advantage of more expensive, newer technologies. Additionally, entrants tend to implement new technologies immediately while incumbent farmers may be slower to adopt due to the costs of replacing older technology and/or the costs of learning new techniques. This can lead to industry turnover as farms unable to remain competitive exit. Across a variety of commodities, new entrants tend to be the largest farms, both in terms of size and scope, suggesting that they are attempting to immediately capture cost advantages (Melhim et al., 2009a/b).

Recent evidence suggests that government payments may also play a role in the consolidation of production. Researchers have found that government payments may provide a means for some farms to grow, leading to more concentrated production at both the individual farm level (O'Donoghue and Whitaker, 2010) and at the ZIP-Code level (Roberts and Key, 2008).

Table 3.4. Average operating profit margin by sales class, 2007

Sales class	Operating profit margin
	Percent
Less than $10,000	-96.1
$10,000 to $49,999	-48.8
$50,000 to $99,999	-12.6
$100,000 to $174,999	1.2
$175,000 to $249,999	8.3
$250,000 to $499,999	16.5
$500,000 to $999,999	22.6
$1,000,000 or more	26.8

Note: Operating profit margin = 100*(net farm income + interest paid – charges for unpaid operator and unpaid labor – charge for management)/gross farm income.

Source: USDA, National Agricultural Statistics Service and Economic Research Service, 2007 Agricultural Resource Management Survey.

What Is the Likely Outcome?

Among the handful of commodities examined in economic studies, the largest livestock farms appear to take advantage of scale economies, which means they will likely continue to grow in the future. However, in most cases, the largest farms are not growing the fastest, suggesting that they are approaching an equilibrium size. This does depend on the sector though. The largest dairy farms, for example, appear to continue to experience unfettered growth while farms producing corn, wheat, apples, and beef all appear to be approaching an optimal size (Melhim et al., 2009a,b). However, it is not clear when, if ever, an equilibrium size will be attained within these sectors. Breakthroughs in technology, veterinary medicine, and management may continue to increase the optimal size of the farm.

Continued shifts in production to larger farms have implications for the use of marketing and production contracts. Currently, nearly two-thirds of all contract production (under marketing or production contracts) takes place on farms with at least $1 million in sales (Hoppe et al., 2008). As production continues to shift to farms of this size, the share of sales produced under contract will also likely expand. Large processors favor long-term relationships with larger producers to ensure a reliable supply of a given commodity at a volume that allows them to operate close to full capacity.

What Is Unlikely to Change?

Some farm structural characteristics, however, are unlikely to change as production shifts to larger farms. Unless radical changes take place within the farm sector, family farms will continue to dominate farming. Roughly 98 percent of all farms and 86 percent of million-dollar farms remain in family hands—where the operator and the operator's relatives own more than 50 percent of the farm business. Family farms' share of total farms has been fairly stable over time in ARMS and earlier data.[5]

Despite the competitive advantages of large farms, a recent ERS study (Hoppe et al., 2010) found that there are still 800,000 small commercial farms, defined here as farms receiving between $10,000 and $249,999 in gross cash farm income (GCFI). These farms account for about 22 percent of U.S. production. Small commercial farms focus on enterprises-like beef, grain and soybeans, and broilers (Hoppe et al., 2010)-that do not necessarily require a full-time commitment of labor, which allows operators to earn offfarm income. The decline in these farms will likely be gradual; some are profitable, and the operators of others are willing to accept losses or place a low value on their labor.

Despite substantial shifts of production to larger farms, individual farms do not have much, if any, market power, nor are they soon likely to. In industries other than farming, most million-dollar firms would still be considered small businesses. For example, the Small Business Administration (SBA) classifies most retail trade and service businesses as small if they have annual revenue below $7 million (and up to $30 million depending on the industry) (U.S. SBA, 2010). By this standard, almost all farms, even the largest ones, would qualify as small businesses, since 90 percent of all million-dollar farms had sales less than $5 million in 2007.[6]

Production Practices and Productivity

Productivity drives growth in agriculture and can be achieved through many different avenues, including the advent of new technologies, managerial techniques, and changes in demand, such as the introduction of new markets. These, in turn, can alter farmers' production practices—or how they manage land, chemicals, and water. And how farmers manage their resources to produce output leads back to how productive farmers are.

Genetically Engineered Crops

Genetically engineered (GE) crops first became commercially available in 1996. These crop varieties included herbicide-tolerant (HT) and insectresistant (Bt) traits that have increased yields due, in part, to fewer crop losses to pests (Sexton and Zilberman, 2011). The Bt traits have lowered the need for pesticide applications and reduced the need for pest scouting. HT traits allow for a single, post-emergent herbicide application. Both traits end up lowering farmers' labor, management, and machinery use. While other traits are being developed—including virus- and fungus-resistant crop strains; varieties resistant to cold or drought; and traits that improve product quality, including increased protein, oil, or vitamin content—the HT and Bt traits are the most important GE crops currently on the market (Fernandez-Cornejo and Caswell, 2006). Widespread adoption of GE crops soon followed their introduction in 1996 (fig. 4.1). While other crop varieties have been developed (e.g., HT canola and sugar beets), corn, cotton, and soybeans make up the bulk of the acres planted to GE crops. By 2007, farmers were using HT soybeans on over 90 percent of all planted soybean acres.

Targeting specific pests with Bt corn and cotton seed varieties has been successful and has led to decreases in insecticide use without insects developing resistance to the new varieties (National Academy of Sciences, 2010). Additionally, farmers adopting the HT varieties for corn, cotton, and soybeans substituted glyphosate for more toxic herbicides. However, the subsequent reliance on glyphosate has led to herbicide-resistant weeds, reducing the effectiveness of the HT varieties. Due to the potential for new pesticide-resistant weeds, the National Academy of Sciences recommends that farmers develop and incorporate more diverse management practices. For example, operators could use herbicide rotations, mixes, and/or application sequences to minimize the spread of HT weeds.

Soil Management Practices

Farmers use soil management practices, such as crop rotation systems and crop residue management (CRM), to control diseases, pests, and weeds; to improve soil quality and production efficiency; and to provide long-term protection of soil and water resources. Greater use of no-till systems has helped meet conservation compliance requirements by reducing erosion and surface runoff and enabling carbon sequestration. The adoption of soil management practices may also provide higher returns. These practices often require, and are facilitated by, technological innovations and higher levels of management skills.

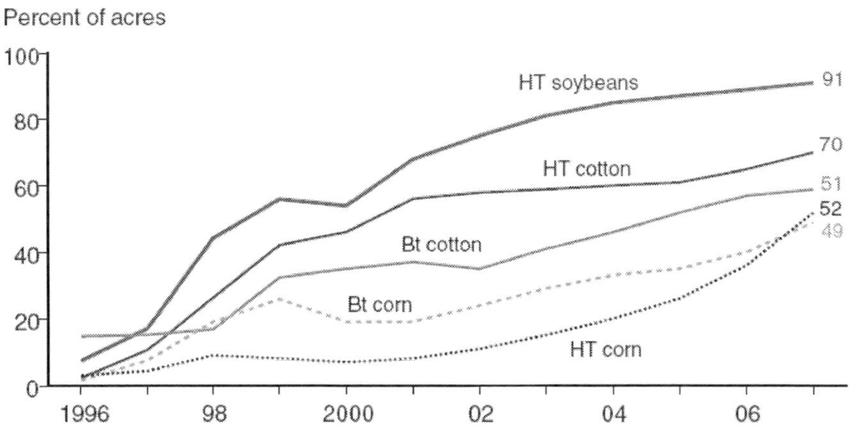

Source: USDA, National Agricultural Statistics Service and Economic Research Service, 1996-1998 Agricultural Resource Management Survey; USDA, NASS *Crop Production,* 1999; and USDA, NASS *Acreage* 2000, 2002, 2004, 2006, and 2008.

Figure 4.1. Adoption of genetically engineered seeds, 1996-2007.

Crop Rotation Systems

The use of crop rotations can help conserve soil, maintain its fertility, protect water quality, and control pests, diseases, harmful insects, and weeds.

Rotating high-residue and/or closely grown crops with row crops can reduce soil losses on erosive soils. Closely grown field grain crops—such as wheat, barley, oats, and hay/forage crops—provide vegetative cover to reduce soil erosion and water runoff while adding organic matter. In addition, these crops help to control broadleaf weeds and may help control weed infestations in subsequent crops.

Beans, lentils, alfalfa, and clover are all leguminous crops that can fixate nitrogen (from the atmosphere) into the soil, making it available for plant growth and reducing the need for commercial nitrogen fertilizers. Legumes in a rotation are most effective in humid and sub-humid climates where they do not deplete subsoil moisture for subsequent crops. Cover crops planted in the fall help reduce erosion from winter and spring storms, hold nutrients that might otherwise be lost, enhance the soil's biological processes, and lengthen periods of active plant growth (to increase nutrient cycling, disease suppression, soil aggregation, and carbon sequestration).

Crop rotations usually result in yields higher than those achieved with continuous cropping under similar conditions. Rotations that add organic matter can improve soil tilth and water-holding capacity, and thus increase crop yields. By alternating a susceptible crop with a non-host crop, crop rotations can help to control a variety of pests by disrupting their life cycles. Additionally, soil microbiology and beneficial insects thrive under crop rotations, helping control disease and other pests, particularly those that attack plant roots.

The diversification inherent in rotations can also provide an economic buffer against fluctuating prices of crops or production inputs and against the vagaries of weather, disease, and pest infestations.

Crop Rotation Use for Major Crops

With the exception of cotton, rotational cropping in some form dominates major crop production in the United States. For example, approximately 60 percent of corn acres and 70 percent of soybean acres in the 10 major producing States used a corn-soybean rotation system in the most recent surveyed year—the 2005 ARMS for corn and the 2006 ARMS for soybeans.

The practice of continuous cropping, whereby the same crop is planted for at least 3 consecutive years, remained constant for soybeans (at roughly 10 percent of all soybean acreage) and wheat (approximately 30 percent) from 1991 through 2007. In contrast, continuous corn planting dropped from 25 percent in 1991 to 17 percent of all corn acreage in 1996 and 2005. For cotton, continuous planting dropped from just over 60 percent of all acreage in 1991 to under 55 percent by 1998, and held steady there through 2004.[1] Wheat, a non-row crop, was planted in a wheat-fallow-wheat sequence to conserve soil moisture more than half of the time in 1991. The increased adoption of no-till, which also serves to conserve soil moisture, led to only 20 percent of all wheat acreage being planted in this rotation by 1998 and 2004.

Crop Residue Management

Crop residue management (CRM) maintains additional crop residue on the soil surface through fewer and/or less intensive tillage operations. CRM is generally cost effective in protecting soil and water resources; can lead to higher returns by reducing fuel, labor, and machinery costs while maintaining or increasing crop yields; and can be quickly implemented to meet conservation compliance requirements. CRM also has fewer resource requirements than structural measures (such as buffer strips, diversions, grass waterways, and terraces) that usually demand higher levels of technical assistance, capital, and contractor skill to install.

CRM systems include reduced tillage, conservation tillage (no-till, ridge-till, and mulch-till), and the use of cover crops and other conservation practices that leave sufficient residue to protect the soil surface from the erosive effects of wind and water (see box, "Crop Residue Management and Tillage System Definitions, " p. 54). In 2004, farmers practiced CRM on roughly 172 million acres (62 percent of planted acres), up from 142 million acres in 1989 (51 percent) (fig. 4.2).

Conservation tillage accounted for 41 percent of U.S. planted crop acreage in 2004 (the most recent data available), compared with 26 percent in 1989. The expansion of conservation tillage has come entirely from the adoption of no-till, which increased from 14 million acres in 1989 to nearly 63 million acres in 2004. The factors driving the expansion in no-till systems fall into three major categories:

- Economic factors that affect farmers' tillage system choices,
- Government policies and programs,
- Technological innovations in chemicals, equipment, and crop genetics.

Crop Residue Management and Tillage System Definitions

Unmanaged, intensive, or conventional till	Crop Residue Management (CRM)			
	Reduced-till	Conservation tillage		
		Mulch-till	Ridge-till	No-till
Moldboard plow or other intensive tillage used	No use of moldboard plow and intensity of tillage reduced	Full-width tillage, but further decrease in till-age intensity	Only the tops of ridges are tilled	No tillage performed since harvest of previous crop
<15% residue cover remaining	15-30% residue cover remaining	30% or greater residue cover remaining on soil surface after planting		

Source: Padgitt et al., 2000.

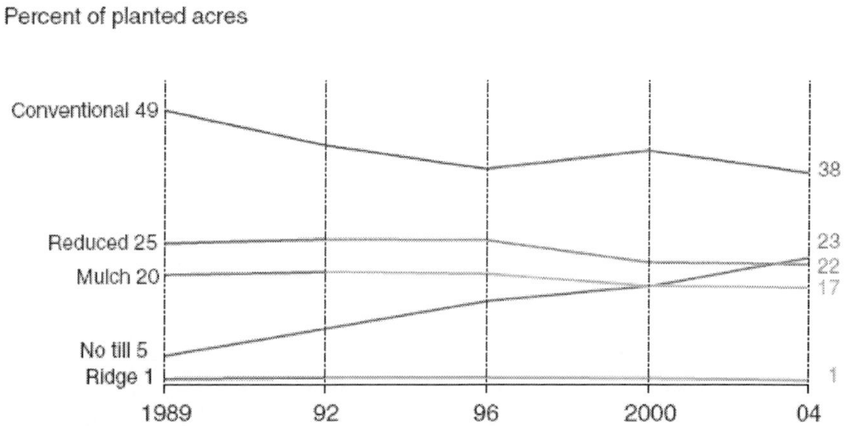

Note: Blue segments show increases.

Source: The Conservation Tillage Information Center, National Crop Residue Management Survey, 1989-2004.

Figure 4.2. U.S. tillage practices, 1989-2004.

Economic Influences

Compared with conventional/intensive tillage, farmers can obtain slightly higher no-till yields on adequately drained soils or on sloping land, particularly with crop rotations. Benefits from improved moisture retention in the root zone usually increase crop yields, especially under dry conditions. In some areas, these benefits enable a change in the cropping pattern to reduce the frequency of moisture-conserving fallow periods.

Choice of tillage system also affects chemical, fuel, labor, and machinery costs. Decreasing the intensity of tillage and/or reducing the number of trips over the field generally lower fuel and labor requirements per acre, extend equipment life, and decrease maintenance costs. Conservation tillage may increase net returns on the entire farming operation even if returns for a particular crop do not increase. For example, a tillage system that requires substantially less labor per acre, though it may slightly reduce returns per acre, could free up labor to service more acres or generate income elsewhere. These cost savings may be offset by increased pest protection and fertilizer costs required to achieve optimal yields when using less intense tillage practices.

Government Policies and Programs

Under the 1985 Food Security Act and subsequent farm legislation, highly erodible land (HEL) used for crops requires implementation of a conservation plan in order to be eligible for USDA farm program benefits. Many conservation plans for HEL involve some combination of conservation tillage/crop residue management, cover crops and/or rotating row crops with less erosive crops such as small grains and hay/pasture, and structural measures to control erosion and reduce water runoff.

Technological Innovations

The development of larger and faster equipment, information and GPS technologies, and precision agriculture tools such as yield monitors and variablerate equipment allows farmers to better match seed, fertilizer, and pesticide applications to areas within a field where they are most needed. The development of genetically engineered seeds and adoption of crop varieties with herbicide tolerance (HT) and/or resistance to specific pests (Bt) has facilitated a shift to less intensive tillage systems, particularly no-till.

The major limitations to adoption of no-till systems for some farmers include additional management skill requirements; expectations of lower crop yields and/or economic returns in some areas or situations (for example, no-till can prevent soils in some areas from warming up and drying out to the extent necessary for proper seed germination by planting time, which can require replanting or switching to a substitute crop); negative attitudes or perceptions; and institutional constraints (for example, obtaining permission from landlords, or even bankers, may prove difficult).

Pesticide Use and Management

Conventional crop farmers rely on pesticides (essentially, toxins spread on the fields) to combat various types of fungi, weeds, diseases, and pests that, if left untreated, could reduce crop yields. These toxins can be hazardous to the environment, to the producers in the fields, and to consumers exposed to pesticide residues. Indeed, markets for organic foods have risen in part due to consumer demand for pesticide-free foods (Greene and Calvin, 1997).

Pesticide Use from 1982 to 2007

We focus on five crops—corn, cotton, fall potatoes, soybeans, and wheat — that account for nearly two-thirds of the pesticides used. Over the 25-year period, pesticide use has changed as shown in the following graphs. (For information about the data sources used, see box "Data on Pesticides,").

From 1982 to 2007, pesticide use on corn, cotton, fall potatoes, soybeans, and wheat (see box, "Data on Pesticides,") decreased slightly, primarily due to drops in herbicide use (the most heavily used pesticide), while fungicides and other pesticide use (such as desiccants and growth regulators) have increased since 1982. The light gray line in figure 4.3 denotes the growth trend of all pesticide use, measured as millions of pounds of active ingredient, indexed to 1982 = 100 for each pesticide type.

Data on Pesticides

In 1982, 1991, 1994, and 1997 (years we have data for total pesticide use), roughly 62 percent of all pesticides used by U.S. agricultural producers were spread on corn, cotton, fall potatoes, soybeans, and wheat (Aspelin, 2003; Lin et al., 1995; Osteen and Szmedra, 1989; Padgitt et al., 2000; and *Agricultural Chemical Usage: Field Crops Summaries*, USDA, NASS, various years). To examine recent trends in pest management practices, we use ARMS data collected from producers of corn (1996-2001, and 2005), cotton (1996-2000, 2003, and 2007), soybeans (1996-2000, 2002, and 2006) and winter wheat (1996-1998, 2000, and 2004).

Estimates by crop for earlier years are from other sources. Estimates by crop for 1982 are from Lin et al. (1995) and Osteen and Szmedra (1989), for 1990 from Padgitt et al. (2000), and for 1991-2007 from *Agricultural Chemical Usage* (USDA, NASS, various years) using estimates of total planted acres (USDA, NASS, *Agricultural Statistics*, 2008). For wheat, separate estimates were made for winter wheat, other spring wheat, and durum wheat, except for 1995-97, when other spring wheat and durum wheat application rates were not separately available.

Total use estimates are compiled from these sources and from estimates of total pesticides used in U.S. agriculture for 1982, 1985, 1988, 1991, 1994, and 1997 from Aspelin (2003). Data by crop are not available for 1983-1989. Data for corn are not available for 2004 and 2006-07 and similarly for cotton (2002, 2004, 2006), fall potatoes (1998, 2000, 2002, 2004, 2006, 2007), soybeans (2003, 2005, 2007), and wheat (1999, 2001, 2003, 2005, 2007). Estimates were imputed for some years using trends in application rates and percentages of planted acres that were treated (area treated) and data on planted acres. When an obvious pattern in the application rate or area treated was apparent, imputed values were estimated using average annual percentage changes. When an obvious pattern was not available, the previous year's value was used, which was often the case for area treated. Crop shares for 1983-1989 are the averages for 1982 and 1990-94.

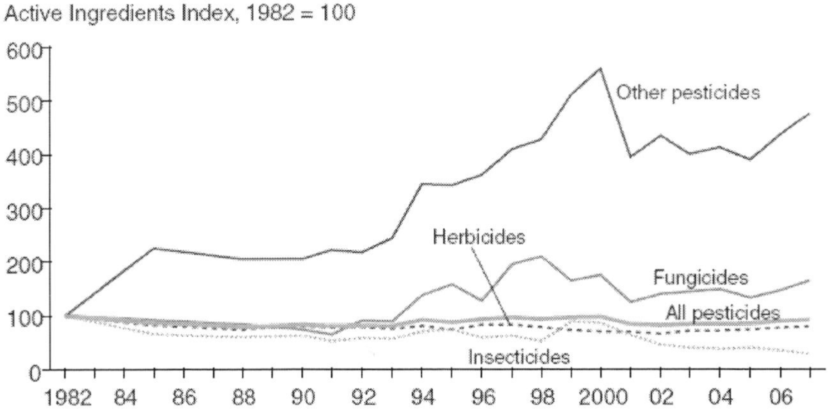

Source: See "Data on Pesticides" box, page 56.

Figure 4.3. Pesticide use, 1982-2007.

Despite drops in herbicide use, herbicides remain the dominant pesticide. Roughly 70 to 80 percent of all active ingredients placed on the five crops examined come from herbicides (fig. 4.4a). Although the share of active ingredients placed on corn crops has dropped, it still receives the highest share, while fall potatoes' share has risen significantly. Meanwhile, cotton's share has fluctuated due to the successful efforts aimed at eradicating the boll weevil (fig. 4.4b).

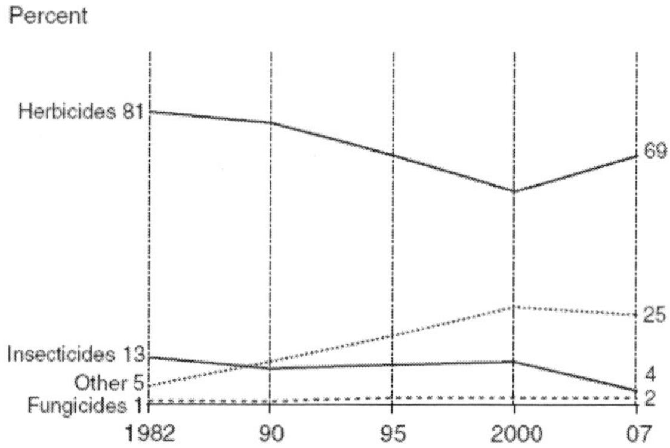

Figure 4.4. (a) Share of total active ingredients used, by pesticide type.

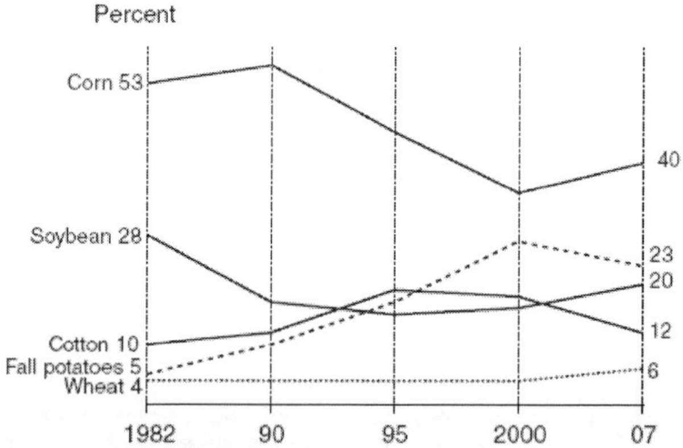

Source: See "Data on Pesticides" box, page 56.

Figure 4.4. (b) Share of total active ingredients used, by crop.

The introduction of genetically engineered seed varieties that allowed plants to tolerate the herbicide glyphosate had different effects on herbicide use, depending on the crop examined. On corn fields, the new seed varieties allowed farmers to replace the more heavily applied butylate and EPTC herbicides with glyphosate, inducing lower application rates of active ingredients. Potato farmers replaced the more heavily applied EPTC with metribuzin, pendimethalin, and rimsulfuron, which also caused a reduction in both application rates and total placement of herbicide active ingredients on potato fields.

The use of glyphosate in soybean and cotton fields, however, had the opposite effect. For these crops, glyphosate replaced herbicides with lower application rates, causing an increase in both application rates and total pounds of active ingredients.

Pesticide Management

Genetically engineered herbicide-tolerant seed varieties allow farmers to scout for weeds and use herbicides only when needed, reducing overall chemical use. Adoption of these seed varieties in corn, cotton, and soybeans over the past 25 years increased the use of post-emergence herbicides and reduced reliance on pre-emergence herbicides for these crops. Farmers of winter wheat, however, came to rely more heavily on pre-emergence herbicides over this period, leading to a decline in scouting for that crop.

Overall, farmers have come to rely less on insecticides. The adoption of genetically engineered seed varieties of corn that express toxins derived from a common soil bacterium, *Bacillus thuringiensis* (Bt)—which kills insect pests such as the European corn borer, the corn earworm, and the corn rootworm — has allowed corn farmers to use fewer insecticides.

Beginning in the 1990s, cotton farmers coordinated efforts to eradicate the boll-weevil, creating a spike in insecticide use. Because these successful efforts are winding down and due to the increased adoption of Bt cotton seed varieties, application rates dropped in the 2000s. Further adoption of the seed varieties, including varieties of corn and cotton that express more than one Bt toxin, will likely lead to continued reductions in insecticide use.

Potato farmers have come to rely more heavily on insecticides with low application rates (especially pyrethoids and imidacloprid), reducing the level of use. Insecticide use on soybean fields was minimal during the 1990s, but picked up in the 2000s, likely due to the introduction of the soybean aphid in northern production areas.

In general, fungicide use increased over 1982-2007; rates of use, however, depended heavily on the crop. Potato farmers increased their use of fungicides with high application rates, such as chlorothalonil and mancozeb.

While soybean farmers had used fungicides sparingly, the introduction of Asian soybean rust into the United States in 2004 changed that. The increased use of fungicides on soybean fields is expected to continue as the pathogen that causes the disease becomes established in the South where it can overwinter (Livingston et al., 2004). In contrast, farmers have reduced their use of fungicides on wheat and cotton crops over 1982-2007.

Nutrient Use and Management

Farmers rely heavily on fertilizer nutrients—especially nitrogen, phosphorus, and potassium—to boost crop yields. However, if the chemicals are not used by the plants, they can pollute air, surface, and groundwater resources.

Management of these nutrients helps determine both crop yields and their environmental impact. Applying more nutrients than the crops require reduces their effectiveness and increases their volatilization into the atmosphere or leaching into groundwater. Application methods and timing are also important. For example, spraying fertilizers onto the fields (versus incorporating them into the soil) increases the likelihood of the chemicals volatizing into the air or being swept away by heavy rains.

Nutrient Use over 1982-2007

Roughly two-thirds of all fertilizer nutrients are spread on corn, cotton, soybeans, and wheat fields. From 1982 to 2007, farmers increased the total tons of nutrients applied to their fields by 9 percent, driven by increases in nitrogen (N) applications, while phosphorus (P) and potassium (K) use dropped (fig. 4.5). (Nutrient application is measured as thousands of tons of applied nutrients, indexed to 1982 = 100 for each nutrient type.) Overall, fertilizer nutrient applications rose at an average annual rate of 0.3 percent.

Corn drives many fertilizer trends since farmers plant the most acres to corn and corn fields receive the highest application rates (MacDonald et al., 2009). Nitrogen makes up more than half of all the nutrients applied on U.S. crops, and its share is increasing due to higher corn plantings (fig. 4.6).

It should be noted, however, that "all other crops" (not shown here) make up a substantial portion of nutrient use. Farmers applied roughly two-thirds of all nutrients on corn, wheat, soybean, and cotton fields (fig. 4.7), but the portion going to all other crops increased from 29 percent in 1982 to 35 percent by 2007.

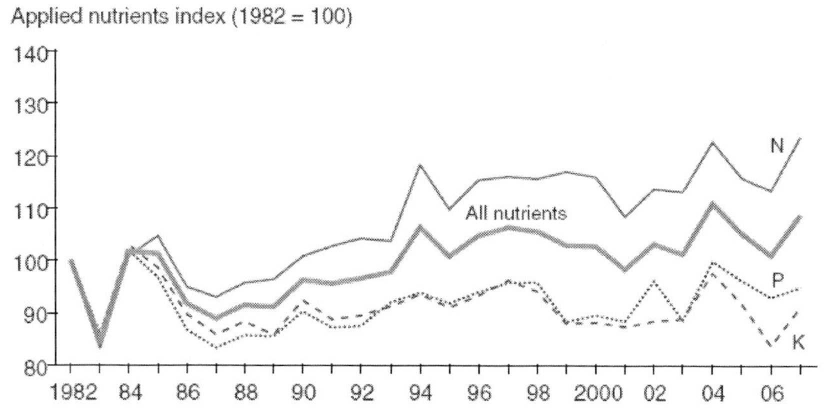

Note: (N) nitrogen, (P) phosphorus, and (K) potassium.

Source: USDA, Economic Research Service, Fertilizer Use and Price (http://www.ers.usda.gov/Data/FertilizerUse/).

Figure 4.5. Nutrient use over 1982-2007.

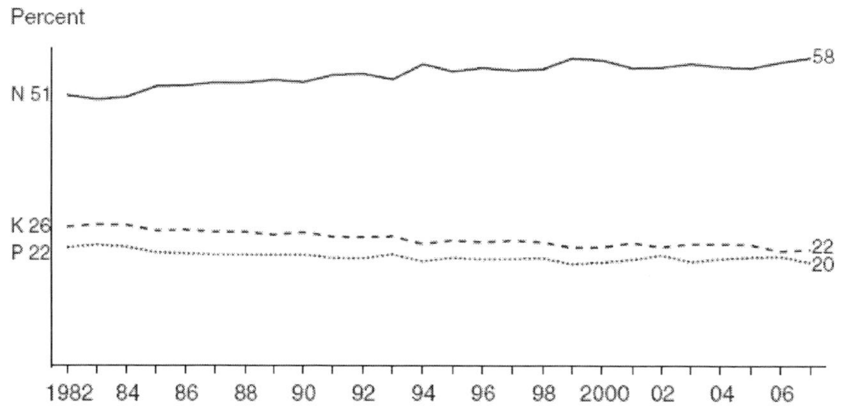

Note: (N) nitrogen, (P) phosphorus, and (K) potassium.

Source: USDA, Economic Research Service, Fertilizer Use and Price (http://www.ers.usda.gov/Data/FertilizerUse/).

Figure 4.6. Share of total tons of nutrients applied to fields, by nutrient type.

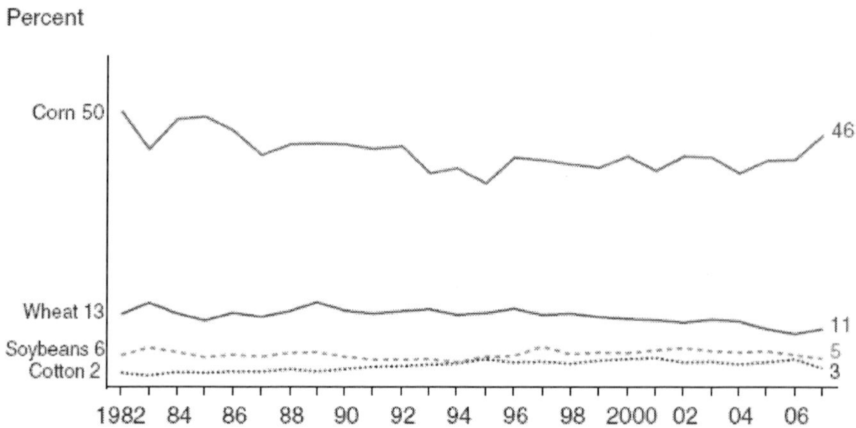

Source: USDA, Economic Research Service, Fertilizer Use and Price (http://www.ers.usda.gov/Data/FertilizerUse/).

Figure 4.7. Share of total tons of nutrients applied to fields, by crop.

Nutrient Management

Nutrient management includes manure management, method of applying the nutrients, the use of various tests to determine nutrient requirements, and nutrient application timing. Shares of planted corn, cotton, soybean, and winter wheat acres receiving manure (organic fertilizer) have not changed markedly since 1996, with corn averaging 14 percent, followed by soybeans (4 percent), cotton (3 percent), and winter wheat (2 percent).

By incorporating fertilizers directly into the soil, farmers help to minimize the amount of nutrients lost to volatilization and runoff. However, this practice can increase the leaching of nutrients into groundwater. Farmers incorporate the majority of chemical fertilizer nutrients applied to corn, cotton, soybeans, and wheat fields. Corn and soybean farmers have increased the share of planted acres that receive chemical N incorporated directly into the soil, while the share of acres planted to cotton and winter wheat that receive incorporated chemical N has fallen moderately (fig. 4.8).

Other management practices include using N-inhibitors, which delay the release of nitrogen, and plant tissue and soil tests to determine the optimum level of additional nutrients (fig. 4.9).

The majority of acres planted to corn, wheat, soybeans, and cotton do not receive N-inhibitors and do not get tested for nutrient content. The scope of these practices has not changed since 1996, and is unlikely to change.

Farm operators can also time the fertilizer application to help reduce nutrient loss through leaching. ARMS data indicate that farmers apply chemical fertilizer nutrients shortly before and/or after planting, when crop demand is greatest, to most planted acres, though such attention to timing has remained static since 1996.

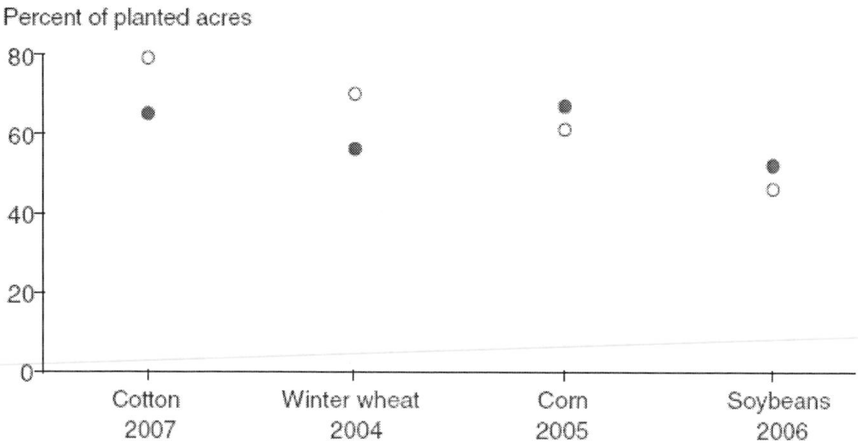

Source: USDA, Economic Research Service and National Agricultural Statistics Service, 1996, 2004, 2005, 2006, and 2007 Agricultural Resource Management Survey, Phase II.

Figure 4.8. Incorporating chemical N into fields, 1996 ( ) and most recent year (•).

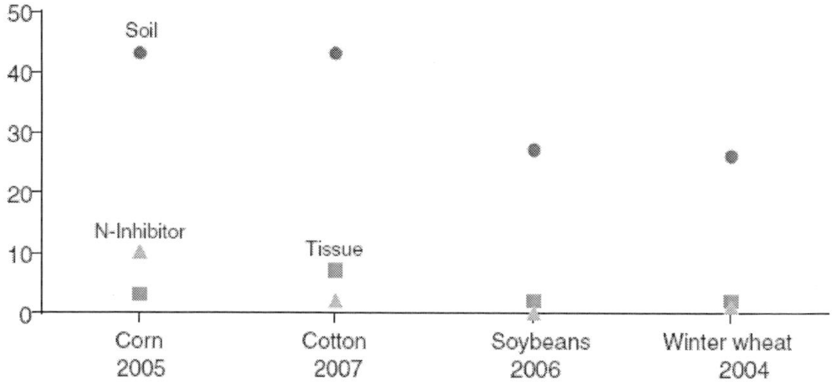

Source: USDA, Economic Research Service and National Agricultural Statistics Service, 1996, 2004, 2005, 2006, and 2007 Agricultural Resource Management Survey, Phase II.

Figure 4.9. Percent of farms that use soil N-inhibitors, plant tissue tests, and soil tests.

Organic Production

In 1990, Congress passed the Organic Foods Production Act (OFPA), requiring the U.S. Department of Agriculture (USDA) to develop national standards to assure consumers that organic products meet consistent, uniform standards. In 2000, the USDA issued a final rule that implemented the National Organic Program (NOP) to govern the USDA organic program. The rule, which took effect in 2002, established national organic standards governing production and handling processes for organic agriculture.

USDA organic regulations cover the scope of crop and livestock production, wild crop, and handling. To increase soil fertility and biodiversity, organic crops must be cultivated using rotation practices, cover crops, and animal and plant material. Weeds, diseases, and pests are generally controlled through physical, mechanical, and biological means, and organic crops are raised without toxic, persistent pesticides, petroleum-based fertilizers, or sewage sludge; any synthetic substances that are applied to the land have to be approved by the regulations. Production standards for livestock require that they be fed 100 percent organic feed, have year-round outdoor access, and not administered antibiotics or growth hormones. Processed products should not contain artificial preservatives, colors, or flavors, and any nonorganic ingredients or processing aids must be approved by the regulations. Comprehensively, use of genetic engineering, ionizing radiation, and sewage sludge are forbidden in organic production and handling.

Certification and Regulation

Under the authority of OFPA and the NOP regulations, operations must be certified by a USDA-accredited certifying agent to represent their products as USDA organic. Certification requires developing an Organic System Plan, which serves as the foundation of the certification process as it documents how a grower or handler plans to meet NOP standards.

The Organic System Plan, specific to the operation type, must enumerate the following elements: farming or handling practices inherent to the operation, such as crop rotations, fertilization, soil conservation, water quality testing, and weed/pest management practices; substances that are used as production or handling inputs; monitoring practices and their frequency, such as soil testing or product quality testing; recordkeeping systems; management practices and physical barriers to prevent commingling with nonorganic products; and any other additional information deemed necessary to evaluate compliance with the regulations.

Full organic status requires a 3-year conversion period, during which no prohibited substances may be applied to the land prior to the harvest of any product that qualifies as organic. Effectively, the producer must abide by organic standards for 3 years without being able to sell the output as organic. This introduces steep set-up costs since organic food costs more to produce.

According to a recent ERS analysis of national dairy and soybean survey data, organic operations absorbed significantly higher total economic costs than their conventional counterparts (Greene et al., 2009). Yet organic products also command significant price premiums at the farm level. Organic milk producers enjoyed an average price premium of $6.69 per hundredweight (44 percent higher than the average price received for non-organic milk) in 2005, which covered most of the additional costs of organic production that year. In 2006, the premium for organic soybeans more than covered the additional costs, making them more profitable than conventional soybeans that year.

Growth over Time

While organically produced agricultural goods make up a small portion of the market, demand has increased for these products, especially under the NOP. In 1980, consumers purchased $470 million (in 2008 dollars) worth of organic food products; they bought nearly $20 billion by 2007 (table 4.1).

Table 4.1. Organic retail sales, 1980-2007

Year	Sales ($ billion (2008))	*(Percent from fruit and veg. in blue)*
1980	0.47	
1990	1.65	
1992	2.36	
1996	4.80	
2002	10.3	42
2007	19.6	37

The Organic Food Program Act of 1990 is implemented, setting a national organic standard and requiring either USDA-accredited State or private organization certification of organic farmers, processors, and handlers. (→ 1990)

Fruit and vegetable sales make up the largest share of organic production and sales. (← 2002)

Source: Natural Food Merchandiser, 1996; and Nutrition Business Journal, 2009.

Market Share

Although only a small segment of the overall agricultural marketplace, organic agriculture has established a foothold in many U.S. farm commodities, with rapid increases in the levels of farmland, livestock numbers, and farm operations (table 4.2)[2]

Table 4.2. U.S. certified organic levels of farmland, livestock numbers, and farm operations (percent of U.S. totals in blue), 1992-2008

	1992	2000	2008	
		Acres		
Cropland	403,400	1,218,905	2,655,382	0.35
Pasture/range	532,050	557,167	2,160,577	0.46
		Number		
Broilers	17,382	1,924,807	9,015,984	0.10
Milk cows	2,265	38,196	249,766	2.7
Beef cows	6,796	13,829	63,680	0.19
Hogs	1,365	1,724	10,111	0.02
Total operations[1]	3,587	6,592	12,941	0.006

In 2008, total certified organic cropland rose to over 2.6 million acres. While growth has been rapid, total certified organic cropland made up only 0.35 percent of total cropland in the United States in 2008.

[1] Does not include subcontracted or exempt organic farm operations.

Source: The Conservation Tillage Information Center, National Crop Residue Management Survey, 1989-2004.

Irrigation

While the United States, as a whole, has abundant freshwater supplies, these supplies are not distributed evenly across the Nation. Agriculture is one of the largest users of fresh water and is by far the largest consumptive user of water (generally, that part of water withdrawals that does not return to the water environment) (Gollehon et al., 2006). In particular, the arid West tends to consume more than half of its renewable water supplies under normal conditions; in drought years, water use often exceeds renewable flow through the use of depletable storage systems, such as aquifers and reservoirs. Together, States in the Pacific, Mountain, and Plains regions accounted for roughly 85 percent of the total agricultural withdrawals of freshwater in 2000 (Gollehon et al., 2006).

Over time, competition for increasingly scarce water resources has intensified, mainly due to population growth, but also due to energy sector growth, ecological and environmental demands, and Native American water-right claims. Climate change projections and an increased interest in biofuels are heightening the tensions surrounding water allocation rights (Schaible et al., 2010).

In response, farmers have implemented water conserving irrigation practices. From 1984 to 2003 (the timeframe in which we have data available to us), total agricultural water use remained relatively stable, dropping by roughly 600,000 acre-feet (or less than 1 percent), while Western irrigation grew to cover an additional 2 million acres between 1984 and 2008 (fig. 4.10).

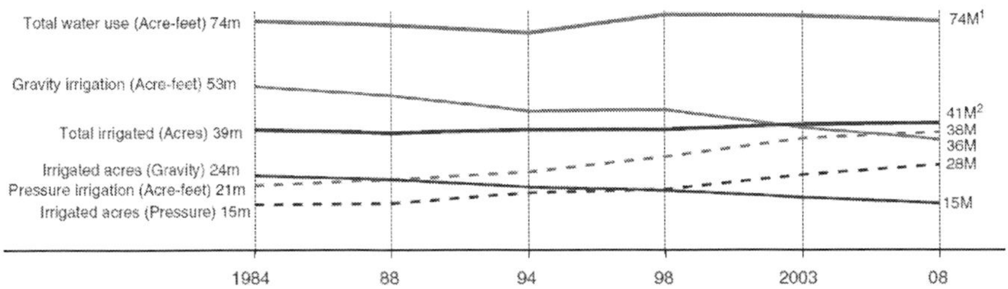

[1] Total water use generally exceeds the sum of the water use of gravity and pressure irrigation. While gravity and pressure make up the bulk of irrigation systems, subirrigation systems are excluded from the analysis.

[2] The sum of gravity and pressure irrigation acres can exceed the total number of acres irrigated. Irrigators may begin with pre-plant irrigations using a gravity system, but then switch to a sprinkler system after crop planting.

Source: USDA, National Agricultural Statistics Service, Farm and Ranch Irrigation Surveys (1984, 1988, 1994, 1998, 2003, and 2008).

Figure 4.10. Trends in water use, 17 Western States, 1984-2008.

Farmers use two main types of irrigation: gravity and pressure irrigation. In gravity irrigation, water uses the path of least resistance from its source to the crops. Pressure irrigation systems rely on pipes to ensure more uniform coverage and reduce losses of water due to evaporation and runoff.

Seventy-one percent of all agricultural water use in the West was applied to crops using gravity systems in 1984. By 2003, operators used gravity systems to apply only 51 percent of all water, with pressure systems applying the other 49 percent. By 2003, the majority of irrigated acres in the West were being irrigated with pressure irrigation systems, many with conserving systems such as drip, low-pressure sprinkler, or Low-Energy Precision Application (LEPA) systems.

Tensions continue to mount concerning the allocation of water resources among competing demands, and the sustainability of Western irrigated agriculture will likely depend on the continued adoption of water conservation practices. These will probably include a continued shift from gravity to pressure systems, and may include more intensive use of infield water-management practices such as soil- or plant-moisture sensing devices, commercial irrigation scheduling services, or computer-based crop-growth simulation models (Schaible et al., 2010).

Productivity in U.S. Agriculture

Changes in farmers' production practices can lead to increases in productivity. At the same time, the spread of new technology can make inputs more effective or allow them to be combined in better ways. U.S. agricultural output has increased significantly, while inputs such as pesticides and water have decreased due to the introduction of GE seeds and the development of pressure irrigation systems.

GE seed adoption has increased yields while helping to reduce labor, management, and machinery use due to reduced pesticide applications and pest scouting (for Bt seeds) and lower herbicide use (for HT seeds). Moreover, GE seeds allow for the increased adoption of no-till practices, which also help decrease farmers' reliance on fertilizers and lower the amount of labor and machinery used otherwise. Scale economies for some commodities (primarily livestock) and organizational and capital innovations in others provide additional sources of productivity gains.

ERS has developed a statistical series—total factor productivity (TFP)—to isolate the effect of changes in technology and related factors from changes in inputs on the growth of agricultural output. Growth in TFP is the primary source of new wealth creation, so trends in agricultural TFP are an important indicator of the longrun performance of the sector.

Agriculture relies more heavily on improvements in technology as a source of growth than almost any other sector of the U.S. economy (table 4.3). Overall growth in industrial production from 1960 to 2004 was almost double that of agriculture.[3] While growth in TFP accounted for 13 percent of the growth in all industrial output over this period, it accounted for 117 percent of the growth in agricultural output. Improvements in agricultural TFP also contributed significantly to the overall productivity growth of the U.S. economy. Even though agriculture accounted for only 1.8 percent of GDP, it accounted for 12.1 percent of all TFP growth in private industry over 1960-2004 (Jorgenson et al., 2006).

From 1982 to 2007, agricultural TFP grew at a rate similar to total output—an aggregation of crop and livestock commodities and related services (fig. 4.11). Total inputs—land, labor, capital, and intermediate inputs like fertilizer, feed, and seed—trended slightly downward due largely to reductions in the amount of labor employed in agriculture. Overall, the amount of crop and livestock output produced per unit of (aggregate) input increased by 35 percent from 1982 to 2007, abetted by productivity growth.

Table 4.3. Sources of growth in agriculture and all industries, 1960-2004

	Agriculture	All U.S. Industries
	Percent	
Average annual growth in output	1.7	3.2
Share of output growth due to growth in:		
Non-labor inputs	11.8	54.1
Labor hours	-34.2	23.7
Labor quality	5.6	8.8
TFP	116.8	13.4
Total	100.0	100.0

Source: Fuglie et al., 2007.

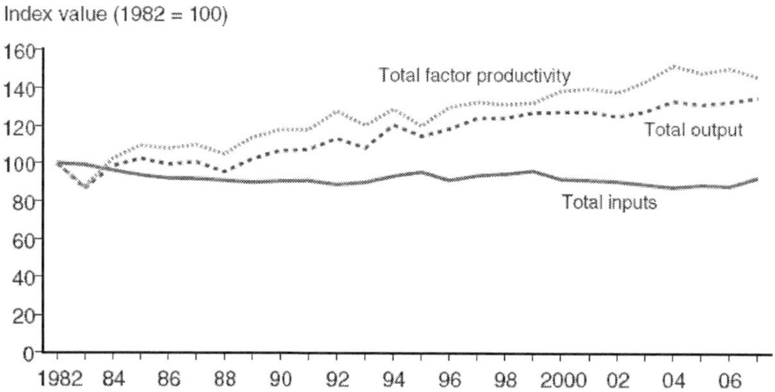

Source: USDA, Economic Research Service, indices of farm output, input, and total factor productivity (http://www.ers.usda.gov/Data/AgProductivity/#datafiles).

Figure 4.11. Indices of aggregate U.S. agricultural output, input, and total factor productivity, 1982-2007.

An Illustration: Productivity Growth in Corn Production, 1982-2007

The aggregate changes in agricultural production and productivity are the sum total effect of myriad changes taking place every year in the way farmers produce individual crops and livestock commodities. These productivity changes occur when new technologies are adopted, for example, or when scale economies are achieved in production.

To illustrate how such changes occur, we focus on corn and some technological changes that have facilitated its production since 1982. Corn accounts for the most value of production in the U.S. farm economy and has arguably experienced more growth in productivity than any other major commodity over the past 70 years (see box, "Earlier Productivity Growth in Corn Production,").

Technology adoption can lead to changes in production practices. Widespread adoption of genetically engineered (GE) corn varieties helps explain the falling share of corn acres treated with insecticides over 1996-2005 (fig. 4.12). Moreover, the adoption of herbicide-resistant strains of corn also contributed to farmers' move toward reduced- or no-till, lowering costs of machinery, fuel, and labor, while reducing soil erosion.

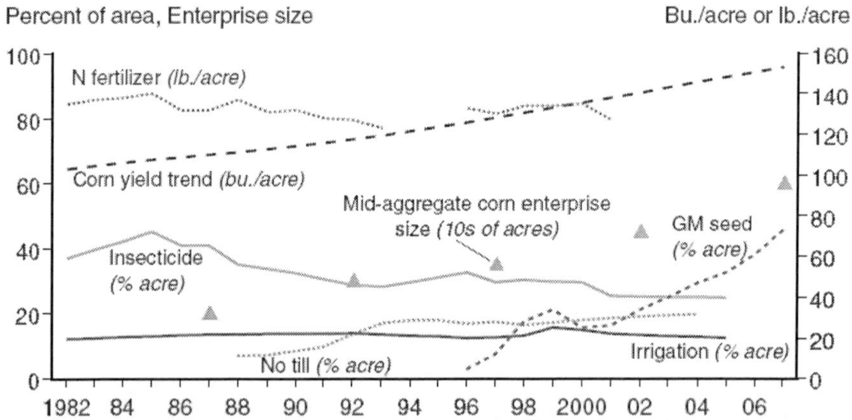

Sources: Fuglie et al. (2007); Census of Agriculture, various years; and USDA, Economic Research Service (2009b, 2009c, and 2010).

Figure 4.12. Changes in GM seed, chemical use, cropping practices, and mid-aggregate enterprise size for corn, 1982-2007.

Other than the increase in GE corn varieties, there were no major changes in production practices for corn after 1982. According to the 2005 ARMS, about 60 percent of U.S. corn fields had been planted to soybeans the previous year, 26 percent previously to corn ("continuous corn"), and the rest to other crops or left fallow. These percentages had not changed much since the first ARMS was conducted in 1996 (USDA, ERS, 2009c). Nitrogen fertilizer application and irrigation practices also remained relatively stable. Some changes in business organization—like the shift in production to partnerships, corporations, and LLCs—could have contributed to productivity increases.

Nevertheless, technology served to increase corn productivity by altering production practices, resulting in the use of fewer resources (land, labor, chemicals, energy, and machinery) to produce a bushel of corn. As a result, production costs dropped. In the long run, with lots of competition (many farmers grow corn) and reduced costs, the prices received for corn tend to decrease as well. Technological advances can therefore not only herald changes in production practices, but also in reduced prices, both in terms of what the farmer receives and, eventually, in what consumers pay.

Recent prices for agricultural commodities have been high, however, reflecting strong demand for commodities for food and for feedstock in the production of biofuels (Parker, 2011). The discussion in the previous paragraph assumes "ceteris paribus," or other factors—including demand—are held constant. Of course, the other factors in reality may not be constant. Reaching the equilibrium described may not occur immediately. Nevertheless, prices may still be lower using new technologies than they would be in their absence, because they allow farmers to increase the supply of agricultural commodities.

Earlier Productivity Growth in Corn Production

Compared with other major crops, corn probably witnessed the most gain in productivity—measured in output per acre—over the past seven decades. Between 1866 and 1940, average corn yield in the United States remained static at about 25 bushels per acre, but since 1940 yield rose steadily, exceeding 150 bushels/acre by 2007.

Farm Mechanization

The first major change to affect corn production in the United States was the conversion from animal and human power to mechanized power for farm operations. The historical evidence on agricultural mechanization in the United States, such as the spread of tractors, refers to all farming operations and not just corn, but the pattern is generally similar for major field crops.

Early gasoline tractors of the 1900s were very large, patterned after the giant steam plows that had preceded them. Improvements between 1910 and 1940 greatly increased their versatility and power and reduced their size. Important innovations prior to World War II included the frameless tractor, pneumatic tires, power take-off, and hydraulic lifting devices for implements (Sundquist et al., 1982). Following World War II, there was a shift from gasoline to diesel-powered tractors and an increase in horsepower. Another important trend was adoption of self-propelled harvesting and threshing equipment and on-farm corn drying technology which reduced post-harvest crop losses.

Mechanization of farm operations did not have a large effect on corn yield but contributed substantially to raising Total Factor Productivity (TFP). Not only did mechanization reduce labor requirements, but it also freed up large amounts of cropland that had previously been used for forage and feed grain production for draft animals. Farms could now devote more land to commercial crop and livestock production. Breaking the link between crop production and feed requirements for draft animals facilitated the regional specialization of commodity production. Regions where corn could not be grown efficiently could now convert that land to other uses; and regions where corn was best suited could now grow corn on lands previously needed for pasture and forage crops.

Yield-Enhancing Innovations

After 1940, the revolution in corn yield materialized, attributable to the development and adoption of a series of innovations involving varietal improvement, fertilization, pest and disease management, the advancement of irrigation, and changes in soil tillage practices.

In the 1930s, hybrid seed began to replace open-pollinated seed and by 1956 had been adopted on over 90 percent of corn acreage in the United States (USDA, NASS, Agricultural Statistics, various issues). After World War II, farm applications of inorganic fertilizers and chemical pesticides rose significantly.

The increased use of fertilizer and pesticides was necessary to realize the higher yield potential in the new hybrid varieties adopted during this period. Nitrogen fertilizer use rose very rapidly from under 20 lb/acre in 1950 to roughly 130 lb/acre by the late 1970s. Herbicide applications reached 95 percent of corn acreage by 1982, while the share of corn under irrigation slowly expanded to reach 13 percent of total acreage by the late 1970s.

Conclusion

This chapter has identified four trends that have implications for the organization of U.S. farming. They are declining land and labor devoted to farming, changing business organization, a shift in production to larger farms, and growth in productivity.

Declining Labor and Land

Two major inputs into farming, labor and land, have declined over the past three decades. The amount of labor dedicated to farming dropped by roughly 30 percent for hired labor and 40 percent for self-employed labor between 1982 and 2007. Meanwhile, the principal operators remaining in farming have grown older, with the share of principal farm operators at least 65 years old increasing from 18 to 30 percent, while the share of those 35 or younger decreased from 16 to 5 percent of all farmers.

While the amount of labor dedicated to farming declined rather steeply, land in agriculture declined more modestly. Land in agricultural use fell from 1.23 billion acres in 1982 to 1.16 billion acres in 2007, or by about 6 percent. Of this 70-million-acre drop, 60 million came from reductions in cropland, which declined from 469 million acres in 1982 to 408 million acres in 2007.

Substantial shifts between pasture and cropland also took place. Roughly 50 million acres transitioned from cropland to pasture or rangeland, while nearly 45 million acres transitioned the other way, from pasture or rangeland to cropland between 1982 and 2007. Additionally, urban areas continued to expand their influence on agricultural land use. Close to 40 million acres of cropland, pasture and rangeland, and forestland were converted to urban uses between 1982 and 2007—conversions that rarely transition back to agricultural purposes.

Organization and Business Arrangements

Commercially oriented farm businesses have come to rely more heavily on contracting. The share of production under marketing or production contracts increased from 28 percent in 1991 to over 37 percent by 2007. Moreover, production has shifted to farm corporations and partnerships.

Together, this means that risks are now spread across a wider set of stakeholders.

In managing risk, farmers have increasingly come to rely upon Federal crop insurance, which grew from 100 million acres covered ($900 million in premiums) in 1989 to over 200 million acres by 2007 (over $6 billion). Government support of farm income has continued to shift from set-aside, price support, and supply control programs to commodity-related and conservation programs not directly tied to farmers' current production decisions. With the introduction of the Conservation Reserve Program in 1986, for example, conservation funding increased from less than $430 million in 1982 (in 2007 dollars) to over $3 billion by 2007.

While smaller farms receive the bulk of conservation program payments (farms with annual sales below $100,000 received almost 60 percent of these payments in 2007), funding has increased for working-land programs (as opposed to land-retirement programs) designed to make farmers' production practices more conservation-oriented. Because production is skewed toward the larger farms, this renewed focus will likely result in larger farms receiving increasing shares of conservation payments, as they do commodity program payments. Farms with sales above $250,000 received roughly 70 percent of commodity program payments in 2007.

Continuing Shifts to Large-Scale Farms

Not only do large farms receive the most support, but they also have a competitive advantage in the marketplace. Farms with sales below $100,000 in 2007 had, on average, negative operating profit margins, while those above $100,000 in sales enjoyed positive average margins that increased with farm size. Large farms, in addition, can adopt new or improved technologies more easily. This suggests a continued shift in production to very large farms. Farms with sales above $1 million already increased their share from 27 to 59 percent of total agricultural production between 1982 and 2007. This also suggests a continued increase in the share of production under production and or marketing contracts since large food processors tend to favor contracting with large farms. Between 1991 and 2007, the use of contracts increased by 9 percentage points, and we can expect further increases going forward.

The number of smaller commercial farms (those with sales of $10,000 to $249,999), in contrast, will likely continue to decline. Very small farms, however, exist independent of the farm economy—the operators of which rely heavily on off-farm income and are less likely to decline in number. Despite probable continued shifts in production to larger farms, farming will likely remain a family-oriented business as the family-farm share of the farm count remained between 97 and 99 percent of all farms from 1988 through 2007.

Productivity

Total factor productivity has continued to increase, growing over 45 percent from 1982 to 2007. This has allowed output to grow substantially with very little change to total input use. Farming depends more on technological improvements for output growth than most other industries and—driven by an increased use of technology—production practices have changed, including the increased adoption of no-till and lower uses of pesticides and insecticides. The increased productivity has kept prices stable for many of the foods produced today.

References

Aakre, Dwight. 2005. *Custom Farm Work Rates on North Dakota Farms, 2004, by North Dakota Farming Regions.* EC-499 (Revised). North Dakota State University Extension Service, Jan.

Aspelin, A.L. 2003. *Pesticide Usage in the United States: Trends During the 20th Century*, CIPM Technical Bulletin 105, Center for Integrated Pest Management, North Carolina State University, Raleigh, NC, February.

Barnard, Charles, Keith Wiebe, and Vince Breneman. 2003. "Urban Influence: Effects on U.S. Farmland Markets and Value", pp. 319-41 in Charles B. Moss and Andrew Schmitz, eds. *Government policy and farmland markets: The maintenance of farmer wealth*, Ames: Iowa State Press.

Bechdol, Elizabeth, Allan Gray, and Brent Gloy. 2010. "Forces Affecting Change in Crop Production Agriculture." *Choices: The Magazine of Food, Farm and Resource Issues.* 25(4).

Bellamy, Donald. 1992. "Educational Attainment of Farm Operators," *Agricultural Income and Finance Situation and Outlook Report.* AFO-45. U.S. Dept. Agr., Econ. Res. Serv. May, pp. 37-39.

Bucks, Brian K., Arthur B. Kennickell, Traci L. Mach, and Kevin B. Moore. 2009. "Changes in U.S. Family Finances from 2004 to 2007: Evidence from the Survey of Consumer Finances," *Federal Reserve Bulletin*, February.

Cochrane, Willard W. 1993. *The Development of American Agriculture: A Historical Analysis*, second edition. Minneapolis: University of Minnesota Press.

Dimitri, Carolyn, Anne Effland, and Neilson Conklin. 2005. *The 20th Century Transformation of U.S. Agriculture and Farm Policy.* EIB-3, U.S. Department of Agriculture, Economic Research Service, June.

Doye, Damona, Darrel Kletke, and Nikki Coe. 2009. *Breeding Livestock Lease Agreements*. AGEC-571. Oklahoma Cooperative Extension Service, September 29.

Fernandez-Cornejo, Jorge and Margriet Caswell. 2006. *The First Decade of Genetically Engineered Crops in the United States*. EIB-11, U.S. Department of Agriculture, Economic Research Service.

Frey, H. Thomas, and Roger Hexem. *Major Uses of Land in the United States: 1982*, AER- 535, U.S. Department of Agriculture, Economic Research Service, 1985.

Fuglie, Keith O. and Paul W. Heisey. 2007. Economic Returns to Public Agricultural Research. Economic Brief No. 10, Economic Research Service, September.

Fuglie, Keith, James M. MacDonald and Eldon Ball. 2007. *Productivity Growth in U.S. Agriculture*. Economic Brief No. 9, Economic Research Service, September.

Gardner, Bruce L. 2002. *American Agriculture in the Twentieth Century: How It Flourished and What It Cost.* Harvard University Press: Cambridge, Massachusetts.

Gollehon, Noel and William Quimby, 2006. "Irrigation Resources and Water Costs," in *Agricultural Resources and Environmental Indicators, 2006 Edition*, eds. Keith Weibe and Noel Gollehon. U.S. Department of Agriculture, Economic Research Service, EIB-16, July.

Greene, Catherine, Carolyn Dimitri, Biing-Hwan Lin, William McBride, Lydia Oberholtzer, and Travis Smith. 2009. *Emerging Issues in the U.S. Organic Industry*. EIB-55. U.S. Dept. of Agriculture, Economic Research Service. June.

Greene, C. and L. Calvin, 1997. "'Organically Grown' Vegetables: U.S. Acreage and Markets Expand During the 1990's," *Vegetables and Specialties Situation and Outlook Report*, VGS-271, U.S. Department of Agriculture, Economic Research Service, April.

Hellerstein, Daniel, 2006. "USDA Land Retirement Programs," in *Agricultural Resources and Environmental Indicators, 2006 Edition*, eds. Keith Weibe and Noel Gollehon. U.S. Department of Agriculture, Economic Research Service, EIB-16, July.

Hoppe, Robert A., Penni Korb, Erik J. O'Donoghue, and David E. Banker. 2007. *Structure and Finances of U.S. Farms: Family Farm Report, 2007 Edition.* EIB-24, U.S. Department of Agriculture, Economic Research Service, June.

Hoppe, Robert A., and David E. Banker. 2010. *Structure and Finances of U.S. Farms: Family Farm Report, 2010 Edition.* EIB-66. U.S. Dept. Agr., Econ. Res. Serv., July.

Hoppe, Robert A., Penni Korb, and David E. Banker. 2008. *Million-Dollar Farms in the New Century.* EIB-42. U.S. Dept. Agr., Econ. Res. Serv., December.

Hoppe, Robert A., James M. MacDonald, and Penni Korb. 2010. *Small Farms in the United States: Persistence Under Pressure.* EIB-63. U.S. Dept. Agr., Econ. Res. Serv.

Janssen, Larry. 1993. "Empirical Analysis of Tenure Patterns and Farm Structure," *Size, Structure, and the Changing Face of American Agriculture.* Arne Hallam, editor. Boulder, CO: Westview Press.

Johnson, James, and Janet E. Perry. 2001. "Business Organization and Arrangements of Farms" in *Structural and Financial Characteristics of U.S. Farms: 2001 Family Farm Report.* Robert A. Hoppe, editor. AIB-768. U.S. Dept. Agr., Econ. Res. Serv., May.

Jorgenson, Dale W., Mun S. Ho, John Samuels and Kevin J. Stiroh. 2006. "The Industry Origins of the American Productivity Resurgence." Working Paper, Department of Economics, Harvard University, October 2.

Keatinge, Robert R., Larry E. Ribstein, Susan Pace Hamill, Michael L. Gravelle, and Sharon Connaughton. 1992. "The Limited Liability Company: A Study of the Emerging Entity," *The Business Lawyer,* Vol. 47, No. 2, pp. 375-460, February.

Key, Nigel and Michael J. Roberts, 2006. "Government Payments and Farm Business Survival." *American Journal of Agricultural Economics.* 88(2): 382-92.

Lin, B.-H., M. Padgitt, L. Bull, H. Delvo, D. Shank, and H. Taylor. 1995. *Pesticide and Fertilizer Use and Trends in U.S. Agriculture,* AER No. 717, Economic Research Service, U.S. Department of Agriculture, Washington, DC, May.

Livingston, M., R. Johansson, S. Daberkow, M. Roberts, M. Ash, and V. Breneman. 2004, *Economic and Policy Implications of Wind-Borne Entry of Asian Soybean Rust into the United States*, Electronic Outlook Report OCS-04D-02, Economic Research Service, Washington, DC, April.

Lund, Philip, and Roger Price. 1998. "The Measurement of Average Farm Price." *Journal of Agricultural Economics.* 49(1): 100-110.

MacDonald, James M., and David E. Banker. 2005. "Agricultural Use of Production and Marketing Contracts," in *Structural and Financial Characteristics of U.S. Farms: 2004 Family Farm Report.* David E. Banker and James M. MacDonald, editors. AIB-797. U.S. Dept. Agr., Econ. Res. Serv., March.

MacDonald, James M., and Penni Korb. 2008. *Agricultural Contracting Update, 2005.* EIB-35. U.S. Department of Agriculture, Economic Research Service, April.

MacDonald, James M., Janet Perry, Mary Ahearn, David Banker, William Chambers, Carolyn Dimitri, Nigel Key, Kenneth Nelson, and Leland Southard. 2004. *Contracts, Markets, and Prices: Organizing the Production and Use of Agricultural Commodities.* AER-837, U.S. Department of Agriculture, Economic Research Service, November.

MacDonald, J.M., M.O. Ribaudo, M.J. Livingston, J. Beckman, W. Huang. 2009, *Manure Use for Fertilizer and for Energy: Report to Congress,* Economic Research Service, U.S. Department of Agriculture, Washington, DC, June.

Melhim, Almuhanad, Erik J. O'Donoghue, and C. Richard Shumway. 2009a. "What Does Initial Farm Size Imply About Growth and Diversification?" *Journal of Agricultural and Applied Economics.* 41(1): 193-206.

Melhim, Almuhanad, Erik J. O'Donoghue, and C. Richard Shumway. 2009b. "Do the Largest Firms Grow and Diversify the Fastest? The Case of U.S. Dairies." *Review of Agricultural Economics.* 31(2): 284-302.

Miller, James J., and Don P. Blayney. 2006. *Dairy Backgrounder.* U.S. Department of Agriculture, Economic Research Service. LDP-M-145-01. July.

Mishra, Ashok K., Ron L. Durst, and Hisham S. El-Osta. 2005. "How Do U.S. Farmers Plan for Retirement?" *Amber Waves*, U.S. Dept. Agr., Econ. Res. Serv., Vol. 3, Issue 2, pp. 12-18.

National Academy of Sciences. 2010. *Impact of Genetically Engineered Crops on Farm Sustainability in the United States (Free Summary).* Committee in the Impact of Biotechnology on Farm-Level Economics and Sustainability; National Research Council.

Natural Foods Merchandiser. 1996. New Hope Communications.

Nickerson, Cynthia, Robert Ebel, Allison Borchers, and Fernando Carriazo. 2011. *Major Uses of Land in the United States, 2007.* Economic

Information Bulletin No. 89, U.S. Department of Agriculture, Economic Research Service, December.

Nutrition Business Journal. 2009. *U.S. organic food sales ($Mil)*—chart 22. Purchased by USDA, Economic Research Service. Penton Media, Inc.

O'Donoghue, Erik J. and James B. Whitaker. 2010. "Do direct payments distort producers' decisions? An examination of the Farm Security and Rural Reinvesment Act of 2002." *Applied Economic Perspectives and Policy.* 32 (1): 170-193.

O'Donoghue, Erik J., Michael J. Roberts, and Nigel Key. 2009a. "Did the Federal Crop Insurance Reform Act Increase Farm Enterprise Diversification?" *Journal of Agricultural Economics.* 60(1): 80-104.

O'Donoghue, Erik J., Robert A. Hoppe, David E. Banker, and Penni Korb. 2009b. *Exploring Alternative Farm Definitions: Implications for Agricultural Statistics and Program Eligibility.* EIB-49. U.S. Dept. Agr., Econ. Res. Serv.

Osteen, C., and P.I. Szmedra. 1989, *Agricultural Pesticide Use Trends and Policy Issues,* Agricultural Economic Report No. 622, Economic Research Service, U.S. Department of Agriculture, Washington, DC, September.

Padgitt, M., D. Newton, R. Penn, and C. Sandretto. 2000, *Production Practices for Major Crops in U.S. Agriculture, 1990-97,* Statistical Bulletin No. 969, Economic Research Service, U.S. Department of Agriculture, Washington, DC, August.

Parker, John, 2011. "The 9 billion-people question: A special report on feeding the world." *The Economist*, February 26.

Peterson, R. Neal, and Nora L. Brooks. 1993. *The Changing Concentration of U.S. Agricultural Production During the 20th Century.* AIB-671. U.S. Dept. Agr., Econ. Res. Serv. July.

Raup, Philip M. 2003. "Disaggregating Farmland Markets," *Government Policy and Farmland Markets: The Maintenance of Farmer Wealth.* Chares B. Moss and Andrew Schmitz, editors. Ames, IA: Iowa State Press.

Roberts, M. J., and N. Key. 2008. "Agricultural Payments and Land Concentration: A Semiparametric Spatial Regression Analysis." *Amer. J. Agr. Econ.*, vol. 90, no. 3 (Aug.); pp. 627-43.

Schaible, Glenn D., C. S. Kim, and Marcel P. Aillery. 2010. "Dynamic Adjustment of Irrigation Technology/Water Management in Western U.S.

Agriculture: Towards a Sustainable Future," Canadian Journal of Agricultural Economics, Special Issue on Western Water, vol. 58, no. 4; pp 433-61.

Sexton, Steve and David Zilberman. 2011. "How Agricultural Biotechnology Boosts Food Supply and Accommodates Biofuels." *NBER Working Paper 16699,* National Bureau of Economic Research, Cambridge, MA.

Sundquist, W. Burt, Kenneth M. Menz, and Catherine F. Neumeyer. 1982. *A Technology Assessment of Commercial Corn Production in the United States.* Agricultural Experiment Station Bulletin 546-1982, University of Minnesota, St. Paul, 1982.

U.S. Department of Agriculture, Agricultural Marketing Service (AMS). 2008. *National Organic Program.* Accessible at: http://www.ams.usda.gov/AMSv1.0/getfile?dDocName=STELDEV3004443&acct=nopgeninfo (as of 8/2/2010). Updated April 2008.

U.S. Department of Agriculture, Economic Research Service. 2009a. Briefing Room: World Trade Organization (WTO). Accessed Feb. 23, 2010 at: http://www.ers.usda.gov/Briefing/WTO/. Updated August 24, 2009a.

U.S. Department of Agriculture, Economic Research Service. 2009b. Agricultural Productivity in the United States, Data Sets. Available at http://www.ers.usda.gov/Data/AgProductivity/. Accessed December, 2009b.

U.S. Department of Agriculture, Economic Research Service. 2009c. Crop Production Practices, Farm Business and Household Survey, Data, Data Sets. Available at http://www.ers.usda.gov/Data/ARMS/app/Crop.aspx. Accessed December, 2009c.

U.S. Department of Agriculture, Economic Research Service.2010. Adoption of Genetically Engineered Crops in the U.S., Data Sets. Available at http:// www.ers.usda.gov/Data/BiotechCrops/. Accessed July, 2010.

U.S. Department of Agriculture, *Forest Service (FS). 2008. U.S. Forest Service Land Areas Report as of September 30, 2007*, http://www.fs.fed.us/land/staff/lar/index.html.

U.S. Department of Agriculture, National Agricultural Statistics Service. 1999. *1997 Census of Agriculture.* Vol. 1, Geographic Area Series; Part 51: United States Summary and State Data. AC97-A-51. March.

U.S. Department of Agriculture, National Agricultural Statistics Service. 2001. *1997 Census of Agriculture.* Vol. 3, Special Studies; Part IV: Agricultural Economics and Land Ownership Survey (1999). AC97-SP-4. December.

U.S. Department of Agriculture, National Agricultural Statistics Service. 2004. *2002 Census of Agriculture.* Vol. 1, Geographic Area Series; Part 51: United States Summary and State Data. AC-02-A-51.

U.S. Department of Agriculture, National Agricultural Statistics Service (NASS). December 2009. *2007 Census of Agriculture* Vol. 1: Part 51, Chapter 1, AC-07-A-51, United States Summary and State Data, a, http://www.agcensus.usda.gov/Publications/2007/Full_Report/usv1.pdf.

U.S. Department of Agriculture, National Agricultural Statistics Service (NASS). 1999. *Crop Production*, Washington, D.C. (Available at: http://usda.mannlib.cornell.edu/MannUsda/viewDocumentInfo.do?documentID=1046.)

U.S. Department of Agriculture, National Agricultural Statistics Service. 2009. Crop Production 2008 Summary, Cr Pr 2-1, (08).

U.S. Department of Agriculture, National Agricultural Statistics Service (NASS). 2010a. *Farms, Land in Farms, and Livestock Operations 2009 Summary.* February.

U.S. Department of Agriculture, National Agricultural Statistics Service (NASS). 2011. *Quick Stats*, Washington, DC.

U.S. Department of Agriculture, National Agricultural Statistics Service. Agricultural Statistics Board. Various years. *Acreage*, Washington, D.C. (Available at: http://usda.mannlib.cornell.edu/MannUsda/viewDocumentInfo.do?documentID=1000/.)

U.S. Department of Agriculture (USDA), National Agricultural Statistics Service. Various years. *Agricultural Chemical Usage: Field Crops Summaries*, Washington, DC.

U.S. Department of Agriculture, National Agricultural Statistics Service. Various years. *Agricultural Statistics.*

U.S. Department of Agriculture, Natural Resources Conservation Service. 2009. *Summary Report: 2007 National Resources Inventory*, Washington, DC, and Center for Survey Statistics and Methodology, Iowa State University, Ames, Iowa. 123 pages. Available online at http://www.nrcs.usda.gov/technical/NRI/2007/2007_NRI_Summary.pdf

U.S. Department of Agriculture, Natural Resources Conservation Service and Iowa State University Statistical Laboratory. *Summary Report: 2007 National Resources Inventory, 2009,* http://www.nrcs.usda.gov/technical/NRI/2007/2007_NRI_Summary.pdf.

U.S. Department of Agriculture, Risk Management Agency. 2010. *About the Risk Management Agency*. Program Aid 1667-02. Revised November.

U.S. Department of Labor, Bureau of Labor Statistics. 2008. *Employment and Earnings*. Vol. 55, No. 1, January.

U.S. Department of the Treasury, Internal Revenue Service. 2008. *Tax Issues for Limited Liability Companies*. Publication 3402. Revised March.

U.S. Small Business Administration. "Choose a Structure," *Small Business Planner*, accessed March 1, 2010. (Available at: http://www.sba.gov/smallbusinessplanner/start/chooseastructure/START_FORMS_OWNERSHIP.html).

Wallandar, Steven, Roger Claassen, and Cynthia Nickerson. 2011. *The Ethanol Decade: An Expansion of U.S. Corn Production, 2000-2009*. EIB-79. U.S. Department of Agriculture, Economic Research Service.

End Note for Introduction

[1] The census of agriculture published selected estimates for 1997 with and without the coverage adjustment, which allows analysts to measure the effect of the adjustment. The adjusted estimate of point farms in 1997 was 416,000, or 50 percent higher than the unadjusted estimate of 277,200.

End Notes for Major Inputs

[1] Work by unpaid workers, such as members of the farm household, amounted to only 4 percent of total farm labor in 2007.

[2] The NRI data used here are from a sample survey and represent the surface area (land and water) of the 48 contiguous States. NRI sample covers all ownership categories, except federally owned land (not including federally owned land). The NRI uses points as sampling units—rather than fields or farms—and the points are revisited over time to provide longitudinal data. NRI data show how land is used and detail the natural resource and environmental conditions of the land (USDA, NRCS, 2009, pp. 1-8). NRI data and ERS's Major Land Use data may give different totals because the former is based on a survey while the latter is compiled from various sources (largely government agencies) and includes federally owned land.

[3] Some of this decline could be due to changes in how NASS accounted for nonresponses in the 2007 Census of Agriculture.

[4] The estimates of farmland subject to urban influence are based on unpublished ERS analysis of NRI data using a 2000 index measuring proximity to population called the "Population-Interaction Zones for Agriculture" (PIZA) codes. The PIZA codes are based on a population-interaction index that provides a cardinal measure of the potential interaction between nearby urban-related population (within 50 miles) and agricultural production activities in each 5-kilometer grid cell in the continental 48 States. For a discussion about

the PIZA codes, see http://www.ers.usda.gov/data/populationinteractionzones/ discussion.htm.

End Notes for Business Organization and Arrangements—and Risk

[1] We classify Federal crop insurance and payments from Government farm programs as business arrangements in this chapter. Classifying Federal crop insurance as a business arrangement is understandable, since it is insurance to mitigate risk (like any other insurance). Classifying the farm programs as business arrangements is less straightforward. Some programs pay farmers when commodity prices fall— or production fails in a natural disaster— thus protecting them from price or production risks. These programs can be viewed as contracts between farmers and the Federal Government to reduce risk. Other programs payments, however, are determined by the farmer's history of production and fixed payment rates set by legislation and do not vary with current economic conditions, such as the level of commodity prices. Nonetheless, these programs are discussed in this chapter, along with the other farm programs.

[2] Most information about the forms of business organization comes from the U.S. Small Business Administration (2010). A new form of business organization, the limited liability company (LLC), is discussed later.

[3] Estimates of the number of family farms are available from ARMS for 1996 through 2007. Estimates for 1988 through 1995 are available from the Farm Costs and Returns Survey (FCRS), the predecessor of ARMS. Note that the definition of "family farm" has changed over the years (see "Family Farms" at: http://www.ers.usda.gov/Briefing/WellBeing/glossary.htm#familyfarm.)

[4] Discussion of the characteristics of contracts is drawn from MacDonald et al. (2004) and MacDonald and Banker (2005).

[5] Consistent contracting data begin in 1991 in USDA surveys.

[6] Owner-operators renting to other farmers accounted for another 13 percent of cropland in the 1999 AELOS. Thus, rented land accounted for 64 percent of all cropland in that year.

[7] A similar program is currently in effect for dairy products, whereby the CCC is authorized to purchase unlimited amounts of butter, nonfat dry milk, and cheese at set prices, effectively setting a price floor for these products (Miller and Blayney, 2006).

[8] Other land retirement programs include the Wetlands Reserve Program (WRP), the Farmable Wetlands Reserve Program (FWP), and the Conservation Reserve Enhancement Program (CREP).

[9] To examine who receives Government payments, we use the 2007 ARMS data, which furnish detailed information on the farms' characteristics—including government payment receipts—as well as farm operator and household characteristics. Since ARMS contacts only farm operators, however, it excludes government payments made to people who do not farm, mainly nonoperator landlords.

[10] For more information on the distribution of government payments, see the report *Structure and Finances of U.S. Farms, Family Farm Report, 2010 Edition* (Hoppe and Banker, 2010).

[11] For example, under the Federal Agricultural Improvement and Reform Act of 1996, Title 1, Section 101 (b) entitled "Purpose," states: "It is the purpose of this act to authorize the use of binding production flexibility contracts between the U.S. and agricultural producers to

support farming certainty and flexibility while ensuring continued compliance with farm conservation and wetland protection requirements." This suggests that these PFC (currently known as DP) payments are targeted toward farm operators (those producing agricultural goods) rather than those who only supply one input (land) to the production process (i.e., landlords). More recent legislation targets certain Government programs to those "actively engaged" in farming. A landlord with a cash rental arrangement is, by definition, not actively engaged in farming. See O'Donoghue et al., 2009b, for a discussion of the term "actively engaged."

End Notes for Specialization and Concentration of Production

[1] The increase in the number of farms in recent years was almost entirely due to the addition of point farms, as discussed later in this chapter.

[2] Estimates from Hoppe et al. (2007) were recalculated to exclude point farms.

[3] An enterprise is the portion of the farm operation dedicated to producing a particular commodity. For example, if a farm grows corn and soybeans, the corn enterprise would include only the acres dedicated to growing corn.

[4] While Lund and Price (1998) coined the term "mid-aggregate point," the measure they use was originally developed and used in both the industrial organization and labor literature in the early part of the 20th century.

[5] For more information, see the ERS briefing room on Farm Household Economics and Well-Being at: http://www.ers.usda.gov/Briefing/WellBeing/glossary.htm#familyfarm

[6] For almost all farm specializations, SBA classifies farms as small businesses if they have annual revenues no more than $750,000. The SBA considers a variety of factors when establishing the cutoff for a particular industry, including: average firm size, the size distribution of firms, startup costs, entry barriers, impact of different cutoffs on SBA programs, and comments from the public.

End Notes for Production Practices and Productivity

[1] In the 1990s, surveys collected data on production practices from each of the major crops annually. By the early 2000s, crops were surveyed intermittently.

[2] Data are from USDA-accredited organic certifiers, and differ, in some cases substantially, from the data collected by NASS in the 2008 Organic Production Survey for several reasons. First, the certifier data include only those farms that certified as organic, and covers all such operations. In contrast, the NASS survey collected data from certified organic producers as well as from farms exempt from the certification process (those with sales below $5,000). Additionally, the 2008 NASS organic farm numbers were coverage adjusted based on 2007 agricultural census responses that required farmers to self-report whether they had any organic production on the farm; in other words, the NASS survey may not have captured all the organic farms despite their effort to adjust estimates for nonresponse. Second, the administrative data underlying the certifier numbers may be based on a broader definition of livestock (e.g. dairy cows) than that used for the NASS survey. Third, the NASS survey reports livestock end-of-year inventories to prevent double counting across multiple

operations (which also may coincide with falling herd numbers); the certifier data cover producer expectations on the dates the producers filed for certification throughout 2008. As a result, the certifier data likely represent an upper bound on the number of certified farms, cropland, and livestock, while the NASS survey results (not shown here) likely represent a lower bound.

[3] We use the 1960-2004 period— rather than 1982-2007—because it is the only period with data for all U.S. industries that are comparable with the data for agriculture.

In: Farming and Farmland in the United States ISBN: 978-1-62257-907-5
Editors: C. Iglehart and G. J. Zsofka

Chapter 2

TRENDS IN U.S. FARMLAND VALUES AND OWNERSHIP*

Cynthia Nickerson, Mitchell Morehart, Todd Kuethe, Jayson Beckman, Jennifer Ifft and Ryan Williams

ABSTRACT

Because farm real estate represents much of the value of U.S. farm sector assets, large swings in farmland values can affect the financial well-being of agricultural producers. This report examines both macroeconomic (interest rates, prices of alternative investments) and parcel-specific (soil quality, government payments, proximity to urban areas) factors that affect farmland values. In the last few years, U.S. farmland values have been supported by strong farm earnings, which have helped the farm sector in many regions to withstand the residential housing downturn. Historically low interest rates are likely a significant contributor to farming's current ability to support higher land values. About 40 percent of U.S. farmland has been rented over the last 25 years. Non-operators (landowners who do not themselves farm) owned 29 percent of land in farms in 2007, though that proportion has declined since 1992.

---

* This is an edited, reformatted and augmented version of United States Department of Agriculture, Economic Research Service Economic Information Bulletin No. 92, dated February 2012.

Photos: USDA, Natural Resources Conservation Service.

Errata

This report, originally released in February 2012, was reissued in March 2012 with the following corrections:

- Starting on the bottom of page 106, a paragraph about the effects of government programs on land values was revised to clarify program attributes.
- Starting on the bottom of page 146, a paragraph was amended to make it clear that the Farm Service Agency compiles data on foreign ownership of farm and forest land, and to delete the final sentence on State regulation of foreign ownership.

SUMMARY

Farm real estate (land and structures) is the major asset on the farm sector balance sheet, accounting for 84 percent of the total value of U.S. farm assets in 2009. Because of this, changes in agricultural land values are a critical barometer of farm sector performance and the financial well-being of agricultural producers. These changes also have implications for a wide range of policy issues, including agricultural competitiveness, industry structure, commodity programs, conservation payments, farmland protection, and local property taxes. In addition to being the largest single investment in a typical farmer's portfolio, farm real estate is the principal source of collateral for farm loans, enabling farm operators to purchase additional farmland and equipment or to finance current operating expenses and meet household needs.

What Is the Issue?

Interest in farmland values has grown in recent years due to dramatic increases in these values—more than 10 percent in 2005 and 2006—and the boom-and-bust cycle evident in the residential land market. Questions abound about the extent to which farmland values are affected by macroeconomic factors such as alternative investments, interest rates, and debt servicing capability. In addition to these farm finance considerations, interest also centers around the extent to which parcel-specific factors—such as soil productivity, proximity to delivery points like grain elevators and highways, and the land's development potential—affect farmland values. Changes in farmland markets have also raised questions about tenure patterns, and the extent to which the benefits of farmland ownership are accruing to owners actively engaged in farming or to non-operating farmland owners.

What Did the Study Find?

Many factors that can affect farmland values and ownership, both macroeconomic and parcel-specific, change simultaneously and they interact in complex ways. Though multivariate analysis is needed to provide definitive answers, examining individual factors one at a time through trend and correlation analyses provides a first step in understanding potential relationships.

Trends in farm incomes, cash rents, and interest rates suggest that farmland values were supported by farm earnings in 2009 and 2010, but there have been periods of imbalances in the recent past. Since 2009, though farmland values have been high, the discounted returns from renting farmland have been higher. Also, in the last 2 years, average income from farming has been more than sufficient to service the debt on farm real estate purchases at current mortgage rates. A "speculative bubble" forming in farmland markets cannot be ruled out, but at a national level, farmland values have been supported of late by fundamental farm factors such as farm earnings. However, over 2005-08 and during 1978-85, farming income was insufficient to service debt on farm real estate purchases. Historically low interest rates are likely a significant contributor to farming's current ability to support higher land values. Increases in interest rates would likely put downward pressure on farmland values.

Strong farm earnings appear to have helped farmland markets withstand the significant downturn in the residential housing market in recent years, though some regions may have experienced modest impacts. In addition to its value in a farming use, farmland near urban areas derives value from its potential to be developed for residential housing and other nonfarm purposes. Significant volatility in urban real estate markets over the last decade has raised questions about the extent to which competing land markets are affecting farmland values. A comparison of rural housing and farmland values during the 2001-04 "boom" years of the housing market reveals that farmland values grew faster than rural housing values in many States. During the housing market downturn (2007-09) that affected all but the Plains and Delta regions, farmland values generally declined more moderately than rural housing values. In 19 States (notably California, Oregon, Washington, and Nevada), farmland values continued to increase even though rural housing values declined. Strong gains in agricultural returns and declining interest rates helped dampen the effects of the housing "bust" on farm real estate values.

Many of the factors that help explain the variation in farmland values across the country are parcel-specific. Farmland with higher soil quality requires fewer production inputs and management time, and land that produces more can enhance expected farming returns and thus farmland values relative to land with poorer soils. In the most rural areas where urban pressures are largely nonexistent, our analysis shows that cropland with better soils is correlated with higher land values. However, a positive correlation between soils and land values is difficult to detect near urban areas because so much of the land's value derives from potential development uses and not farm factors.

The correlation between government payments and cropland values varies regionally and by payment type. Government agricultural program payments increase income from agricultural production, and when they do so in a consistent way, the expectation of future payments may be capitalized into the value of farmland. Although farmland values generally increase with insurance premium subsidies, land values are inconsistent for land with different levels of direct and countercyclical program (DCP) payments. Also, the ratio of DCP and insurance premium payments to land values varies regionally, a consequence of what crops are—and have historically been—grown and whether they are eligible for program payments.

Non-operating landowners play a significant role in U.S. agriculture. Ownership status affects whether the benefits and risks of owning farmland accrue to active farmers or non-operating landowners. Three of the top four regions in terms of land in agriculture (Northern and Southern Plains and the Corn Belt) have non-operating owners owning more than 30 percent of the land. Non-operators owned 29 percent of all land in farms in 2007, and they owned 77 percent of farmland that is rented. Non-operators tend to be older, less likely to live on the farm, and less likely to participate in conservation programs. Despite recent increases in foreign ownership of forest land, as of February 2009, only 1.7 percent of privately owned land in farms or forest, or 22.8 million acres, was owned by foreigners.

How Was the Study Conducted?

This report uses trend and correlation analyses to identify general relationships between farmland values and both macroeconomic and parcel-specific factors. The analyses use cropland and pastureland value estimates developed by USDA's National Agricultural Statistics Service (NASS) and data from NASS' June Area Survey, in conjunction with data on a variety of factors—such as proximity to urban areas, soil quality, and irrigation status—that are likely to affect farmland values. Our analyses on farmland ownership trends largely use data from NASS' 2007 Census of Agriculture, NASS' 1999 and 1988 Agricultural Economics and Land Ownership Surveys, and ERS-NASS Agricultural and Resource Management Surveys. Because trends in farmland markets are determined by complex interactions among many potential influences, these trend analyses can help inform the design of detailed statistical analyses that would more definitively identify those factors that are significantly affecting trends in farmland values and ownership.

INTRODUCTION

U.S. farmland values and the factors influencing changes in those values have long been of considerable interest, and the subject of a great deal of economic research. Farm real estate (land and structures) is the major asset on the farm sector balance sheet, and accounted for 84 percent of total U.S. farm assets in 2009. Farm real estate is the principal source of collateral for farm loans, enabling farm operators to finance the purchase of additional farmland

and equipment or to meet current operating expenses. Many farm households rely on landholdings as a retirement fund, so farm real estate is an investment strategy as well. Because of this, changes in agricultural land values are a critical barometer of farm sector performance and the financial well-being of agricultural producers. Changes in land values also have implications for a wide range of policy issues, including agricultural competitiveness, industry structure, commodity programs, conservation payments, and farmland protection.

About 922 million acres, or 40 percent of the Nation's land base, was in farms in 2007 (latest year for which data are available). Most farmland is in the Midwest, but it exists in all States so fluctuations in farmland markets can be both a regional and national concern (fig. 1).

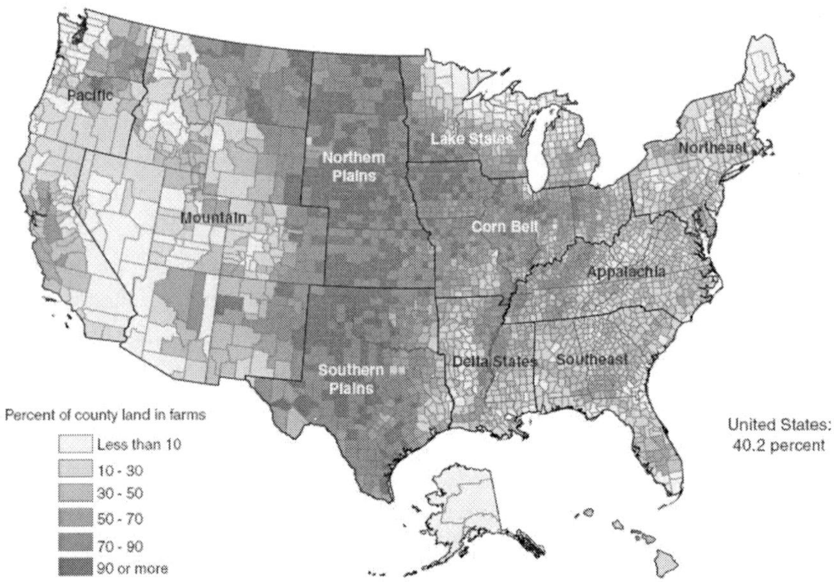

Source: USDA, National Agricultural Statistics Service.

Figure 1. Acres of land in farms, 2007.

Because farmland is such a significant asset in the farm sector, farmers, lenders, researchers and policymakers have long been interested in the trends in, and determinants of, farmland values. This interest has grown in recent years due to dramatic increases in farmland values; rapid changes in commodity prices arising, in part, from renewable energy policy; and volatility in other land markets. For example, residential housing prices escalated in

2005-06 and then dropped precipitously, coincident with a significant increase in commodity prices. These changes have raised questions about the extent to which farmland markets respond to changes in fundamental farm-related factors, such as commodity prices and farm income, and the extent to which they are subject to risks and changes in other land markets.

Sharply rising farmland values also evoke reminders of the trends that preceded the farm crisis of the 1980s. Farmland values rose throughout much of the post-World War II period; from 1969 to 1978, farmland prices increased 73 percent as agricultural producers responded to high returns and Federal policies that increased incentives for investing in agriculture (Barnett, 2010). In 1980, farmland prices began to decline in response to Federal monetary policy that raised interest rates to combat inflation. The rapid escalation in interest rates and energy prices precipitated a farm financial crisis that led to farm bankruptcies and bank failures throughout the 1980s (Duncan, 2008).

Farmland values are determined by factors that affect the land market as a whole, as well as factors that affect the value of individual parcels relative to each other. Examining factors at both of these scales is important in understanding how and why farmland values are changing. Macroeconomic factors behind farmland values—such as expected returns from agricultural uses, alternative investment opportunities, interest rates, and use of debt financing—and other sector-level factors help inform whether farmland continues to be attractive as an investment and whether farmland values appear to be supported by fundamental farm factors. A micro-level perspective reveals how farmland values vary spatially due to variation in parcel-level characteristics like soil productivity, the land's proximity to markets, eligibility for farm program payments, and opportunities for nonfarm income. In this report, we examine both macro and micro aspects of farmland markets.

Farmland Values: Trends over Time, by Region, and Land Use

Since the farm crisis of the mid-1980s, farm real estate values (including land and buildings) have been rising in both nominal and real (i.e., inflation-adjusted) terms (fig. 2). Between 1994 and 2004, real values increased between 2 and 4 percent annually. In 2005 and 2006, values jumped 16 percent and 11 percent, before slowing to 6-7 percent in 2007 and 2008.

Farmland Values Vary by Region and by Land Use Type

Average land values for the Nation mask wide regional variation. Cropland values (the land value excluding buildings) exceeded $5,000 per acre in 2007 in the Northeast and Pacific regions, where high-value crops—fruits, vegetables, and orchards—are grown and development pressures are high (fig. 3). Cropland values in 2007 were only about $1,000 per acre in the Northern Plains, an area dominated by wheat and other row crop farms.

Also, though national average cropland values changed little between 1999 and 2004, cropland values in some regions were more variable. For example, in the Pacific region, cropland values increased between 2 and 4 percent over 1999-2003, but decreased 16 percent between 2003 and 2004.

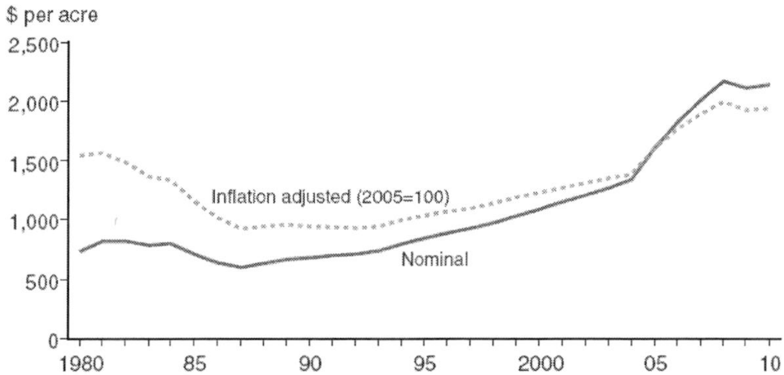

Source: USDA, National Agricultural Statistics Service.

Figure 2. Average farm real estate values, 1980-2010.

Average farmland values also obscure differences by type of land use. Historically, crop and pasture land prices did not diverge significantly because pasture was essential for feeding farm animals, which were an important part of the farming enterprise (Raup, 2003). More recently, when measured, nationally cropland values have exceeded pasture values due primarily to higher returns associated with cropland. Cropland enjoys a substantial premium over pastureland values in most regions (fig. 3), especially in the Pacific States where vineyards command higher land prices (Reynolds and Johnston, 2003).

However, cropland value premiums have declined over the past 10 years in every region except the Northeast. Though premiums exceeded 20 percent at some point within the last 10 years, by 2008, cropland values in the

Appalachian and Southern Plains regions exceeded average pastureland values by only 3-4 percent.

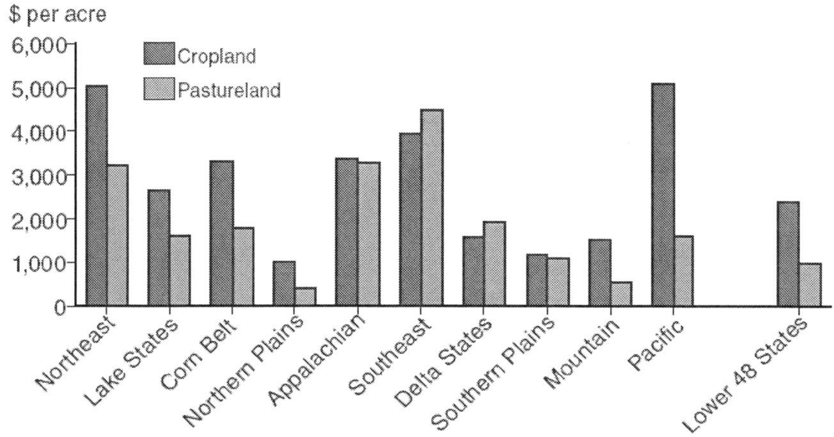

Source: USDA, ERS analysis of National Agricultural Statistics Service June Area Survey data.

Figure 3. Average farmland value by land use and region, as of 2007 (in 2005 $).

Table 1. Percentage of premium cropland over pastureland values (in 2005 $)

Region	1999	2000	2001	2002	2003	2004	2005	2006	2007	2008
					Percent					
Northeast	31	35	37	37	37	52	56	62	56	61
Lake States	133	128	116	113	104	97	80	69	65	67
Corn Belt	136	122	114	105	104	102	98	90	85	93
Northern Plains	196	189	183	186	184	176	167	153	148	148
Appalachian	14	16	20	19	18	21	5	2	2	3
Southeast	15	16	10	4	2	14	-14	-12	-12	-13
Delta States	4	3	7	4	0	-2	-10	-11	-18	-17
Southern Plains	28	30	28	32	37	44	21	13	7	4
Mountain	315	310	318	319	318	254	231	191	178	171
Pacific	360	396	404	399	390	218	262	247	215	193
Lower 48 States	180	184	183	181	178	178	178	140	146	153

Source: USDA, ERS analysis of National Agricultural Statistics Service Agricultural Land Values and Rents Final Estimates.

In the Southern Plains, nonfarm factors such as recreational income from hunting leases are contributing to increasing farmland values (Guiling et al., 2008; Henderson and Moore, 2006; Pope, 1985). (Nonfarm income streams that accrue to pastureland would contribute to declining cropland premiums.) In the Southeast and Delta regions, average pasture values rose more quickly in the last 10 years and have exceeded average cropland values since 2005 (table 1). In Georgia, Flanders et al. (2004) observe that more pastureland than cropland is near urban areas, drawing stronger demand for housing or recreational development.

Performance of Farmland as an Investment

Recent sharp increases in farmland values raise questions about whether farmland prices have reached levels that cannot be sustainably supported by expected agricultural returns. Examining how farmland values change relative to farmland rents, whether farm income is sufficient to service real estate debt, and how farmland compares to alternative investments indicate how farmland has been performing as an investment. These comparisons also reveal that the recent high farmland prices are not occurring under the same conditions that contributed to the farm crisis of the 1980s, at least for the farm sector as a whole.

At the most basic level, capital asset pricing theory posits that farmland values are derived from the expected stream of returns that can be earned into the future by having the land in an agricultural use.[1] Farm incomes and land values were closely linked in the first half of the 20th century, but have become less so since (Shalit and Schmidt, 1982), with little correlation at the national level in recent decades (fig. 4).[2,3]

Rent-to-Value Ratios Declined over the Last 50 Years

The farmland rent-to-value (RTV) ratio, calculated as the cash rent per acre divided by the land value per acre, is a proxy for how quickly an asset will pay for itself.[4] The roughly 45-year trend reveals a decreasing RTV ratio (fig. 5); if agricultural rents were the sole source of returns from farmland, the farmland would have paid for itself in about 14 years in 1951, but would take more than 33 years in 2007. However, in many places, nonagricultural returns can be earned from the land, such as returns from developing the land (Livanis et al., 2006). Decreasing RTV ratios are consistent with the growing importance of nonagricultural factors in determining land values. Indeed, Flanders et al. (2004) find that in Georgia, nonagricultural factors have a stronger influence on land values than cash rents.

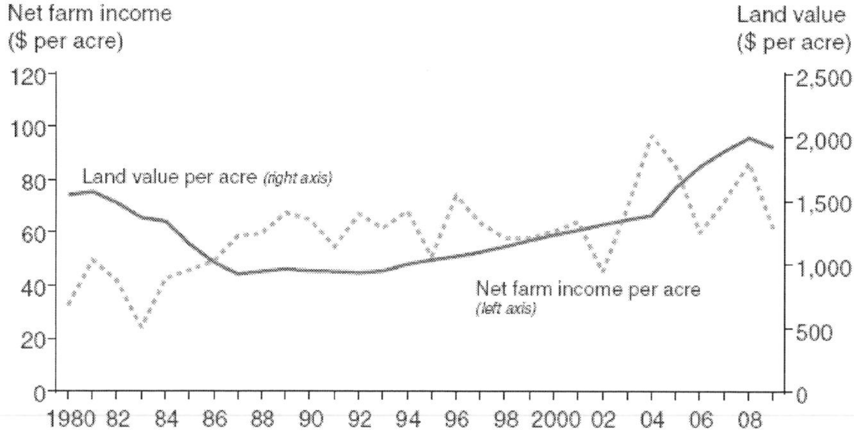

Source: USDA, ERS Farm Income Accounts data, available at http://www.ers.usda.gov/data/FarmIncome/Finfidmuxls.htm

Figure 4. Land values and farm sector net income, 1980-2009 (in 2005 $).

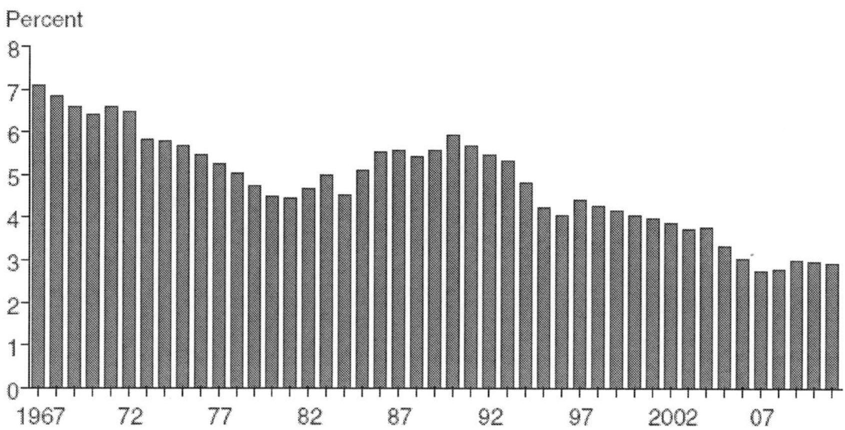

Sources: For 1997-2011, USDA-NASS Quickstats (http://quickstats.nass.usda.gov/). For earlier data, see USDA-ERS cash rent and Agricultural Land Values Survey data.

Figure 5. Cropland rent-to-value, 1967-2011.

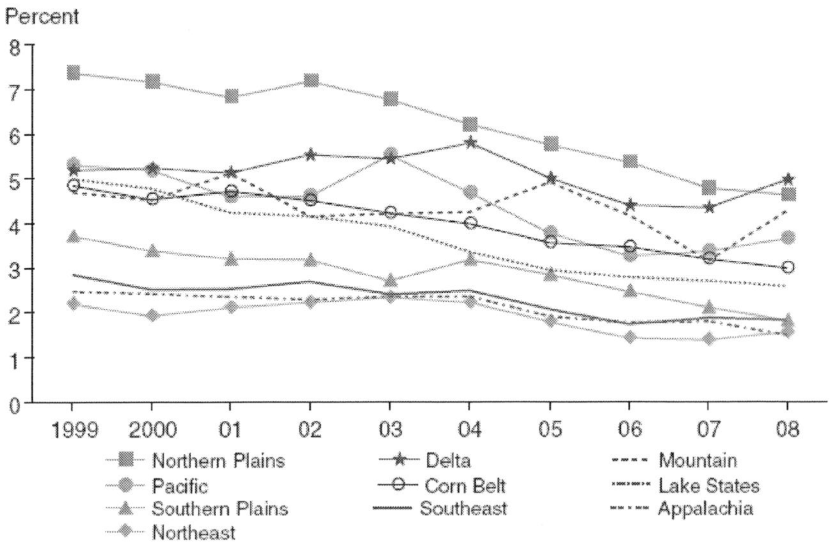

[1] Ratio of per-acre cash rent for cropland to per-acre value of cropland in the June Area Survey, including both irrigated and non-irrigated cropland.

Source: USDA, ERS analysis of National Agricultural Statistics Service June Area Survey data.

Figure 6. Cropland rent-to-value by region, 1999-2008[1].

Survey data collected by USDA's National Agricultural Statistics Service (see box, "June Area Survey Data") show that since 1999, almost every USDA region has seen a decrease in RTV ratios (fig. 6). The Lake States and Southern Plains experienced the largest percentage decreases in RTV ratios, at 48.3 percent and 51.2 percent. The Delta and Mountain regions had the lowest percentage decreases in RTV ratios between 1999 and 2008, at 4.2 percent and 8.8 percent. The Delta, Mountain, and Pacific regions experienced the most volatility in RTV ratios.[5] The regions with the highest RTV ratios are the Northern Plains, Delta, Mountain, and Pacific regions.

Declining rent-to-value ratios in many regions may be due to the influence of nonfarm factors on land values, as RTV ratios are generally lower near urban areas (fig. 7). In regions such as the Corn Belt, that are heavily dominated by agriculture,[6] other factors may be playing a role in declining RTV ratios, such as increased recreational (hunting) values and long-term rental arrangements with rental prices that adjust slowly to rapidly appreciating farmland values (fig. 8). Indeed, studies have found that farmland prices are more volatile than rents (Falk, 1991; Featherstone and Baker, 1987).

Rising incomes in the nonfarm rural economy in recent years may also have contributed to increases in farmland values (e.g., Henderson and Novack, 2005), with a more limited impact on rents.

June Area Survey (JAS) Data

USDA's NASS survey each June asks farmers to report their estimates of per acre cropland, pastureland and total real estate value (including building value), as well as cropland rents. The JAS is based on a probability area frame with an annual sample of about 10,800 segments of approximately 1 square mile. Land values and cash rents are collected from all farmers having operations within the sampled segments (NASS, 2009a). The Sample tool in the Spatial Analyst extension of ArcGIS 9.3 was used to extract the values of other data in GRID format to the JAS segment points, so that associations can be made between the JAS data and characteristics of other data, listed in Appendix A: Data Inputs. A key objective of the JAS spatial data development process is to capture the time-series lineage of the survey at unique spatial locations to generate a single set of points in a spatial dataset that represent survey locations over time.

JAS segments are surveyed longitudinally over several consecutive years, so by design, each point represents at least one survey year. We developed the spatial dataset for the years 1999 through 2008. MS Access was used to develop a table of unique sample points, defined by latitude and longitude fields. Each of these 31,242 unique spatial locations was assigned an identifier. A separate lookup table is used to join the year, State, and segment identifier to the unique location identifier, creating a one-to-many type join. Any extractions of the gridded data to the JAS segment points yielded a table including the spatial location identifier tabbed with any values intersected in the sample process. The lookup table provides the full linkage between the extracted grid data and the JAS tabular data.

Affordability of Farm Real Estate Has Varied over Time

As farmland values surge and rent-to-value ratios decline, questions arise about the affordability of farmland. Figure 9 shows U.S. total farm real estate values from 1970, and the maximum farm real estate value that current farming income can "cash flow" (service debt on) in each year.[7] Although affordability will fluctuate above and below real estate values due to price and yield variation, during two periods in particular—1978-85 and 2005-08—income from farming alone was insufficient to service the debt on farm real estate purchases.

The low levels of affordability indicate that nonfarm factors likely had a large role in determining farmland values during these periods.[8] In other periods, such as before 1974 and between 1986 and 2004, farming incomes were more aligned with the value of land.

When farmland values are high and affordability is low, some farmers might sell land to capture the capital gains benefits arising from appreciated land values.

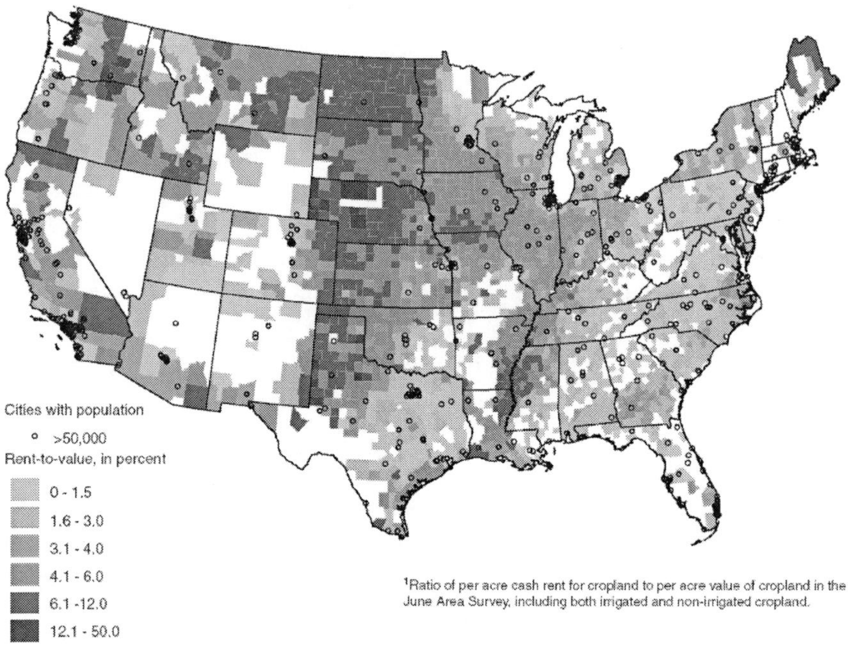

Note: The nonshaded areas either did not have any farms included in the JAS or had insufficient observations due to low levels of agricultural production.

Source: USDA, ERS analysis of National Agricultural Statistics Service June Area Survey data.

Figure 7. County average rent-to-value ratios[1], 1999-2008.

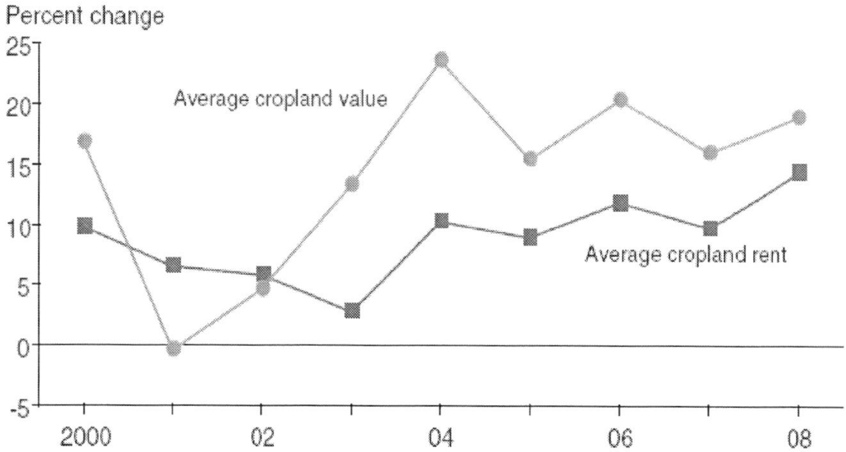

Source: USDA, ERS analysis of National Agricultural Statistics Service June Area Survey data.

Figure 8. Annual changes in per-acre average cropland value and rent in the Corn Belt, 2000-08.

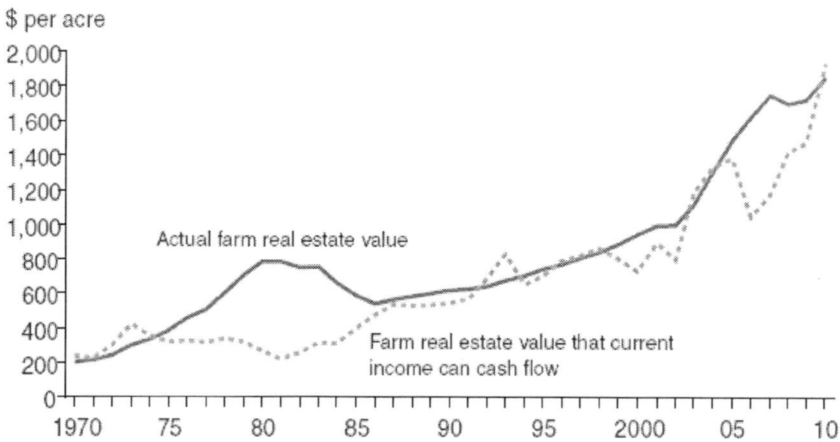

Sources: USDA, ERS Farm Balance Sheet data, http://www.ers.usda.gov/Briefing/FarmIncome/Data/Constant-dollar-table.xls.

Figure 9. Current value of farm real estate versus maximum affordable value $ per acre.

Yet, other farmland owners will not necessarily react by immediately selling their land. Farmland markets have historically been very thin, with some estimates indicating about 0.5 percent of U.S. farmland is sold annually.[9] Farmland owners may hold onto land if the family has held it for a long time or because they derive other nonfinancial (lifestyle) benefits from owning the land. Indeed, studies of farmland market activity reveal that sales of farmland are often forced due to death or retirement rather than a result of affordability levels (Raup, 2003).

Nonetheless, for farm operators who owned the land they operated in 2007, about 53 percent of land was acquired at arm's length from an unrelated seller, about 25 percent of acres owned were inherited, and 19 percent were purchased from relatives.[10] For purchases from relatives at least, evidence suggests the buyers negotiate terms that are more favorable, which can help in periods of high farmland prices or low levels of affordability. Robison et al. (2002) find that farmland was sold to relatives at a discount when compared to land sold to nonrelatives. Debt-servicing difficulties and high real-estate transaction costs have led to increased demand for leasing rather than owning land (Sherrick and Barry, 2003).

The current environment of low interest rates has two positive effects on agricultural real estate values (Schnitkey, 2010). First, for those who have to purchase land with debt capital, it lowers the total cost of purchasing land. The effect of the recent declines in interest rates is evident in figure 9 by the sharply upward trend in the amount of land value that can be cash-flowed with falling farm mortgage rates.

Indeed, farm mortgage interest rates have been at record lows in very recent years, and farmland purchasers are increasingly using debt financing. Agricultural and Resource Management Survey (ARMS) data reveal that land purchases by farm operators in 2008 and 2009 were financed predominantly by credit (74 and 65 percent, respectively) and, in 2009, over half of farm real estate transactions involved debt financed for 75 percent or more of the purchase price.

Second, interest rates represent returns on competing fixed investments, and when they are low, farmland looks more attractive as an investment alternative. Coupled with increases in crop prices and improved agricultural credit conditions, some regions experienced double-digit increases in farmland values in 2010 (Oppendahl, 2010; Briggeman and Akers, 2010).

Low interest rates also aid those who are buying farmland for other uses of the land (see box, "Trends in Farmland Purchases").

Trends in Farmland Purchases

The sharp increases in farmland values in recent years have sparked an interest in the characteristics and intent of farmland buyers. In the 10th District of the Federal Reserve[1], farmland appears to have been purchased by nonfarmers much more for recreation than for development or residential purposes. Purchases of farmland for development in this region have been decreasing since 2002, while purchase of farmland for recreation weakened in 2008. Henderson and Akers (2009) report that this decline in demand has led to a stabilization of land prices after several years of increasing prices. Based on the Iowa 2009 Farmland Value Survey, about 23 percent of land sales were to investors, a decrease from previous years. Likewise, more farmers were reported to buy land (Duffy, 2009).

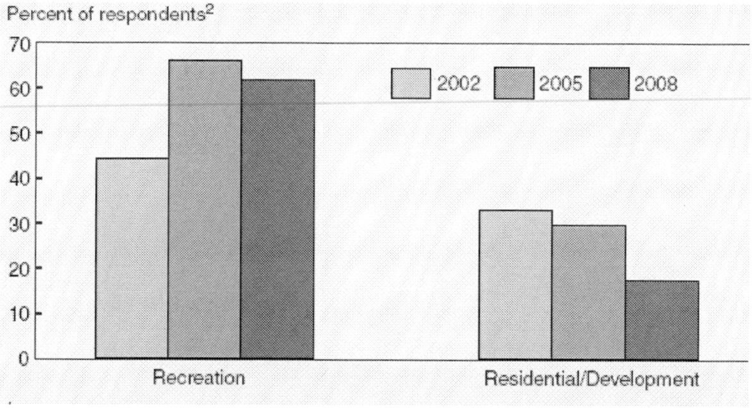

[1]Includes Colorado, Kansas, Nebraska, Oklahoma, Wyoming and portions of western Missouri and northern New Mexico.

[2]Respondents were asked the most common reasons for farmland purchases by individuals other than farmers. Respondents could choose more than one response, and therefore percentages will not sum to 100.

Source: Henderson and Akers (2009).

Reason for farmland purchases by nonfarmers (10th Federal Reserve District).[1]

[1] Includes Colorado, Kansas, Nebraska, Oklahoma, Wyoming, and portions of western Missouri and northern New Mexico.

Farmland Values Were Supported by Farm Earnings in 2009-10

From an investment perspective, capitalized values are used to quantify impacts of both cash rent and interest rate changes on land values. A capitalized value represents the estimated discounted value of all future cash flows to farmland. The capitalized value is calculated by dividing cash rent (a measure of cash flow generated from agricultural production) by a discount factor, typically an appropriate interest rate. When farm real estate values divided by the capitalized value (price-to-value ratio) exceed 1, farmland values are not supported by the stream of cash flows the farmland could earn. During 2004-08, farmland values were not justified based on farm earnings alone (fig. 10).[11] Subsequently, market signals indicate that farmland was somewhat undervalued at current interest rates, so it is not surprising that farmland values increased in 2009-10.

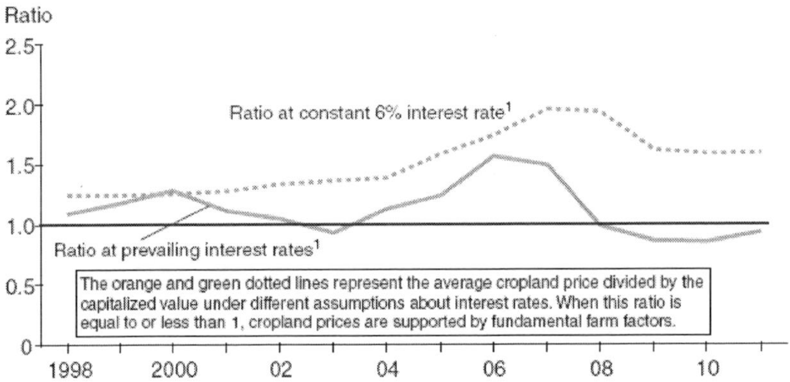

2011 forecast.

[1] Ratios are calculated as the average cropland value per acre divided by the capitalized values. Capitalized value = per-acre cropland cash rent / interest rate on 10-year note. The dotted line represents the same ratio, with capitalized value calculated using a constant 6% interest rate.

Sources: USDA, ERS analysis of National Agricultural Statistics Service farmland value and rent data.

Figure 10. Cropland price-to-value ratios, 1998-2011.

The separate effects of changes in the interest rate are also evident in figure 10. In 2000, current rates were equal to the long-term average Treasury note rate of 6 percent. The dotted green line shows changes in the cropland priceto-value ratio when the interest rate is held constant at this 6-percent average. Since 2000, this price-to-value ratio has declined below what it would have been had interest rates remained at the long-term average. In 2010, the price-tovalue ratio was about 0.9 based on current rates. If interest rates were to jump to the long-term average, the price-to-value ratio would sharply increase to over 1.5, signaling that land values would not be supported by the stream of rents the land could earn at historical interest rates. Farmland markets could subsequently react with sharp reductions in farmland values. Although in the 1980s interest rates did increase abruptly (and farmland values declined dramatically), such abrupt changes are not common. Gradual changes in interest rates have a more muted effect on farmland markets because expectations change and market participants have more time to adjust.

Some predict that stable farm incomes and interest rates will keep farmland values stable, although increasing volatility in the agricultural markets could lead to reductions in farmland values (Henderson and Akers, 2009; Schnitkey, 2010). In addition to limited changes in interest rates, recent trends in several other macroeconomic factors differ from patterns during the farm crisis of the 1980s. In the early 1980s, increasing crop supplies, a strong dollar, and weakened export demand contributed to rapid declines in farmland values. At present, the demand for crops for food and as an energy source is keeping commodity stocks tight and export demand for corn and other commodities strong. Also, some predict that the dollar is likely to remain weak for some time (Henderson, 2010). However, some caution that the size of the U.S. trade deficit increases the risk of investing in farmland. If other countries become less willing to cover the U.S. deficit, interest rates could climb and debt-financed real estate values could fall (Raup, 2003).

Farmland Has Historically Been a Stable Investment and Has Become More Attractive Relative to Other Investments in Recent Years

Farmland serves both as a productive asset and capital investment. Like any capital investment, an appreciation in farmland price carries important (unrealized) capital gains. To examine the returns to investing in farmland relative to other investments, we estimate returns to farmland versus the S&P 500 and gold. Returns are measured as the percentage change in value from year to year. The percentage change is a simple measure that reflects the value of holding an asset for one additional period, which should reflect the expected income generated from the asset.

The expected return of an asset is a function of the investment horizon. Assets that are held for many decades can yield returns that are quite different than assets held for shorter time periods. To account for this, we present the return of a $1,000 investment in (1) farmland, (2) a stock fund tied to the S&P 500, and (3) gold over three time horizons: a return in 2008 to purchases in 1970, 1980, and 1990. Our comparisons are based on real returns using the GDP price deflator (with a base year of 2007), so that the annual returns represent an appreciation in asset value without the influence of inflation. In addition, the initial investment of $1,000 is expressed in real terms in 2007 dollars. For example, in real terms, a $1,000 initial investment in farmland is equivalent to $4,630 in 1970; $2,183 in 1980; and $1,376 in 1990. Given an initial investment in farm real estate made in 1970, both gold and the S&P 500 composite index outperform farmland returns (fig. 11a).

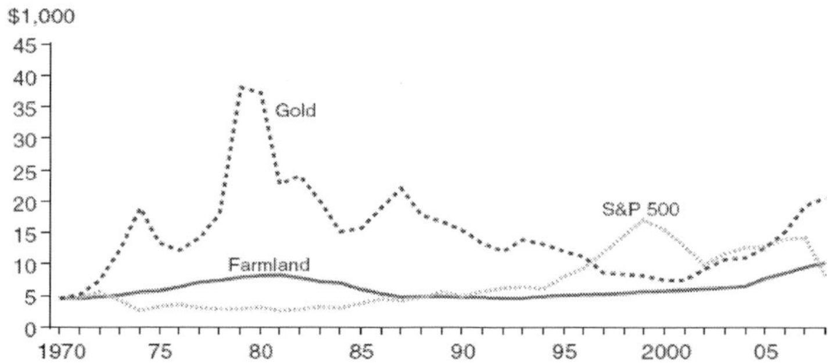

Sources: Farmland returns are derived from the USDA, National Agricultural Statistics Service June Area Survey estimates of average value per acre of all farms and buildings in the continental United States.The S&P 500 and gold prices are based on end-of-year settlement prices.

Figure 11a. Returns to farm real estate versus stocks and gold, 1970-2008.

Although gold exhibits the highest degree of variability, the average annual rate of return is 7 percent, followed by the S&P at 3.2 percent and farmland at 2.3 percent. Over a shorter investment horizon—when farm real estate is purchased in 1980 rather than 1970—the S&P 500 is the favorable investment, with an average annual appreciation rate of 4.9 percent (fig. 11b). Farmland returns appreciate 1 percent per year while gold exhibits a net loss of less than 1 percent per year, on average. The low returns to farmland are not surprising given the farm crisis of the early 1980s, during which farmland values were greatly diminished. Between 1984 and 1985, farmland values declined by a record 14 percent. Over an even shorter investment horizon, the same investment in 1990 nets similar returns for farmland (4.4 percent) and the S&P 500 (4.6 percent per year) (fig. 11c). Although gold declined in value throughout the 1990s, returns rebounded in 2002, yielding an average annual return of 2.5 percent. Farmland values exhibited a relatively stable increase compared to the other investment alternatives, across all three investment horizons.

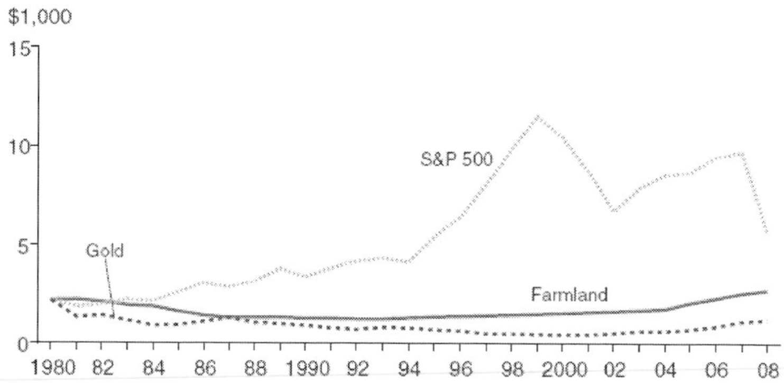

Sources: Farmland returns are derived from the USDA, National Agricultural Statistics Service June Area Survey estimates of average value per acre of all farms and buildings in the continental United States. The S&P 500 and gold prices are based on end-of-year settlement prices.

Figure 11 b. Returns to farm real estate versus stocks and gold, 1980-2008.

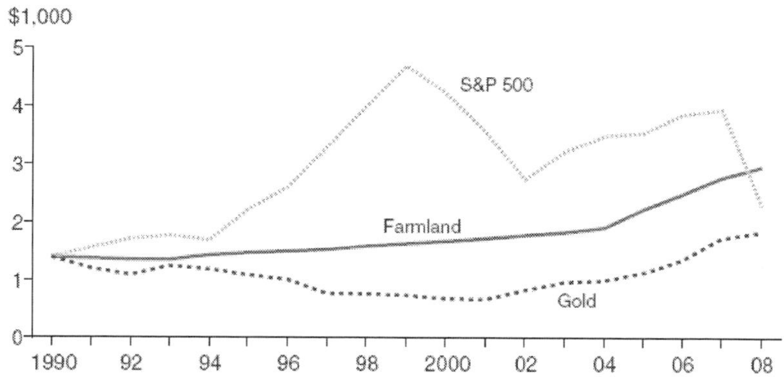

Sources: Farmland returns are derived from the USDA, National Agricultural Statistics Service June Area Survey estimates of average value per acre of all farms and buildings in the continental United States. The S&P 500 and gold prices are based on end-of-year settlement prices.

Figure 11 c. Returns to farm real estate versus stocks and bonds, 1990-2008.

PARCEL-SPECIFIC FACTORS AFFECTING FARMLAND VALUES

The previous section demonstrates how macroeconomic factors can affect farmland values. Yet, factors that are unique to each farm parcel also affect the parcel's ability to generate returns, and these differences give rise to spatial variation in farmland values. Indeed, an indirect approach to measuring the impacts of net returns or rents on farmland value—one that is often employed in the economic literature—is to examine the relative importance of parcel-specific characteristics in explaining variation in farmland values across the country.[12]

Farmland is capable of generating several types of returns, depending on its use. In some areas, farmland value may be derived solely from its ability to support agricultural production, so parcel-specific factors such as soil quality, access to market terminals, and government payments are important determinants of land values. However, farmland close to urban areas may also generate nonfarm returns from the option to develop land for residential, commercial, or industrial uses. Parcel-specific factors related to development potential include distance from population concentrations and access to roads. Another nonfarm factor affecting farmland values is the amenity value of land. Amenity values arise from the scenic views, varied topography, mild temperatures, and other factors that attract people to rural areas and make certain locations particularly desirable. Surveys of real estate brokers (Duffy, 2009) reveal—and economic studies confirm—that both farm-related and nonfarm-related factors are likely to be important in explaining variation in farmland prices. Regardless of the type of returns land generates, the relative importance of such spatially explicit factors is likely to vary across location and time.

The June Area Survey conducted by USDA's National Agricultural Statistics Service (NASS) provides a unique opportunity to examine relationships between farmland values and some of the most commonly considered, spatially explicit determinants of farmland values (see box, "June Area Survey Data" on p. 7). National analysis using spatially explicit data is a complex task: it requires data sources that use consistent definitions for the entire country and significant computational resources. Our analyses using these data demonstrate the importance of using spatially explicit data to understand local influences on farmland values. We construct estimates of land values as the weighted average of the values reported for the tracts within each segment (the sampling unit). Each segment contains a geocode, allowing segment-level values to be linked with a variety of spatially explicit data that may affect farmland values.

Farm-Related Factors Affecting Farmland Values

More Productive Soils Are Correlated with Higher Cropland Values

Farmland with better soil quality can improve yields and reduce production inputs and management time, which can enhance farming returns and thus farmland values. However, the contribution of better quality soils to farm-land values can vary depending on the proximity of the land to urban areas (Palmquist and Danielson, 1989). Some studies have found that demand for land in urban uses is a significant component of land values in areas near large towns and cities (Chicoine 1981; Shi et al., 1997; Livanis et al., 2006; Hardie et al., 2001). In such cases, the effect of urban influence can be so strong that it masks the relationships between farmland values and characteristics that contribute to the agricultural earnings potential of the land.

In demonstrating the soil quality/land value relationship, we control for the effects of urban-related demand by limiting the JAS data to observations that were at least 25 miles from the borders of cities of at least 50,000 people. Our measure of soil quality is based on a "land capability class" measure that represents the soils with the fewest limitations for agricultural uses (see Appendix A). For this subset of more "rural" data points, cropland values were significantly and positively correlated with better soil quality in some regions.[13] These findings are consistent with empirical studies that found a positive (and statistically significant) relationship between soil quality measures and farmland values (Huang et al., 2006; Xu et al., 1993; Palmquist and Danielson, 1989; Chavas and Shumway, 1982).

The effect of soil quality on cropland strengthened as we limited the data to even more rural parcels (those at least 40 miles from population concentrations of at least 50,000). However, patterns vary substantially across regions (fig. 12). In 2008, there was a positive correlation between our soil quality measure and cropland values in the Corn Belt, Lake States, and North Dakota (in the Northern Plains region), but a negative or insignificant correlation in the Appalachian region.

Also, we found that relationships between soil quality and land values varied by irrigation status. In the JAS data, respondents in eastern States were not asked to provide separate estimates of cropland values for irrigated and non-irrigated land, whereas respondents in the western States—where irrigation is more prevalent—were asked to do so. In most regions where cropland value data were available by irrigation status, irrigated land values exceeded average non-irrigated land values, as expected (fig. 12).

Also, the correlation between soil quality and irrigated land values was stronger (the slope of the irrigated land value line is steeper) than between soil quality and non-irrigated land values.

Our analysis of the correlation between soil productivity and farmland values demonstrates the importance of examining parcel-specific factor/land value relationships at a very disaggregated level. For example, when analyzing the relationship at a national level, a positive correlation between better soils and cropland values was not evident over 2005-08.[14] The analysis also demonstrates that controlling for parcel-specific factors other than urban influence is particularly important in some regions.

A study by Palmquist and Danielson (1989) in North Carolina that did so found strong and significant negative relationships between farmland values and erosion potential or wetness of the land (two characteristics that determine the Land Capability Class), contrary to the finding above for the Appalachian region. The more intuitive findings in that study—that better soils increase farmland values— may also be attributable to the more detailed measures of erosivity and wetness that the authors were able to construct for their limited study area.

Land Value Effects of Proximity to Market Channels Vary by Location

Because transportation costs increase with distance, the proximity of land to important points of delivery for crops can affect net returns and land values. Again, using spatially explicit data is important when examining these relationships, at least when points of delivery are unevenly distributed. For example, grain elevators are heavily concentrated in several Corn Belt States and along the Mississippi River. For farms in these regions, there is little variation in the distances between cropland parcels and a grain elevator, making it difficult to ascertain a relationship (fig. 13).

In the Northern Plains where there is more variation in farm distances to elevators, the expected relationship between proximity and cropland values is more evident. The cropland value-proximity relationship is even more pronounced when the data are limited to a single State (North Dakota) (fig. 14).

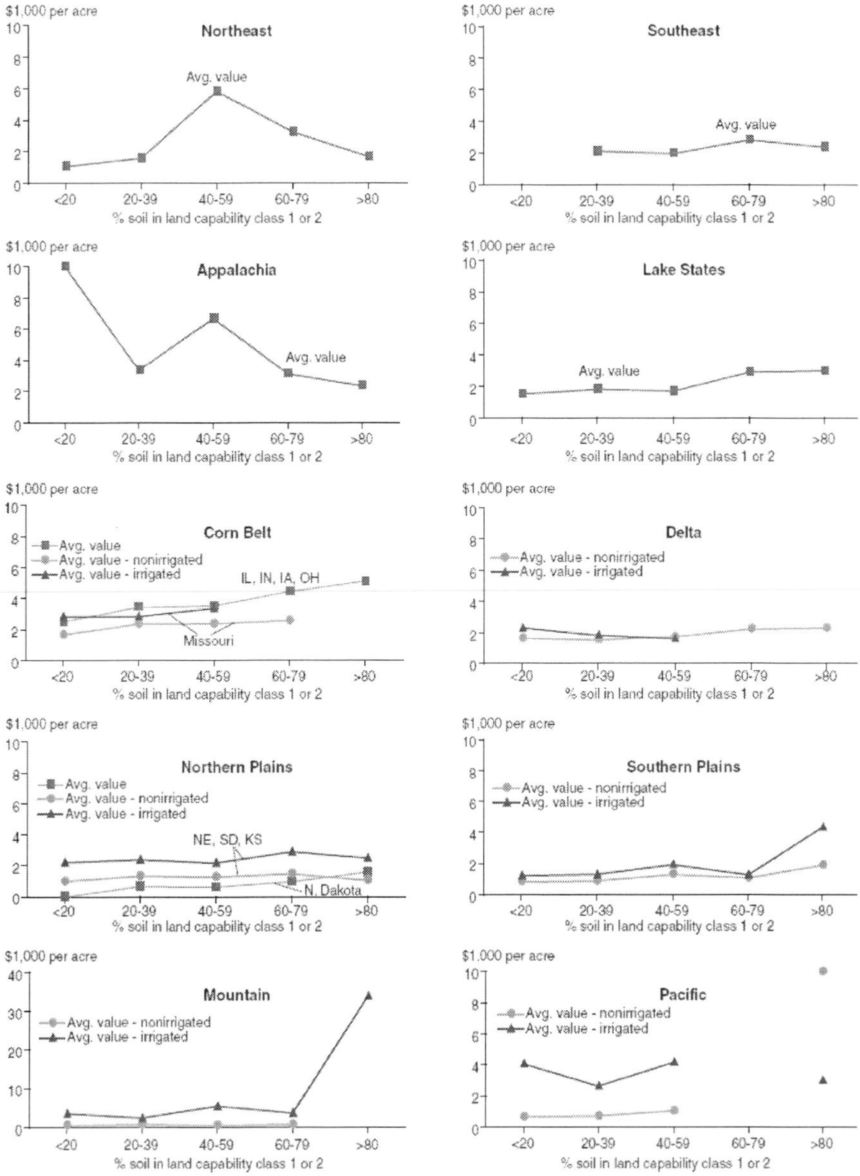

Source: USDA, ERS analysis of National Agricultural Statistics Service, June Area Survey data.

Figure 12. Average per-acre cropland values and share of county soils in land capability class 1 or 2 by region, 2008.

Proximity to ethanol plants also illustrates location effects. In recent years, Federal energy policies supporting the production of biofuels through renewable fuel standards and tax credits have increased the demand for corn, which has elevated corn and other commodity prices. In addition to spurring general crop price increases, ethanol production reduces local transportation costs to terminal markets and provides a competing source of demand for grain. Research has found evidence of higher corn prices up to 68 miles from ethanol plants (McNew and Griffith, 2005). Other research—based on a survey of agricultural lenders in the seven States covered by the Federal Reserve Bank of Kansas City—found that cropland adjacent to an ethanol plant can be valued more than $150 per acre higher than cropland more than 50 miles away, with proximity to multiple ethanol plants producing a higher premium than proximity to a single plant (Henderson and Gloy, 2009).

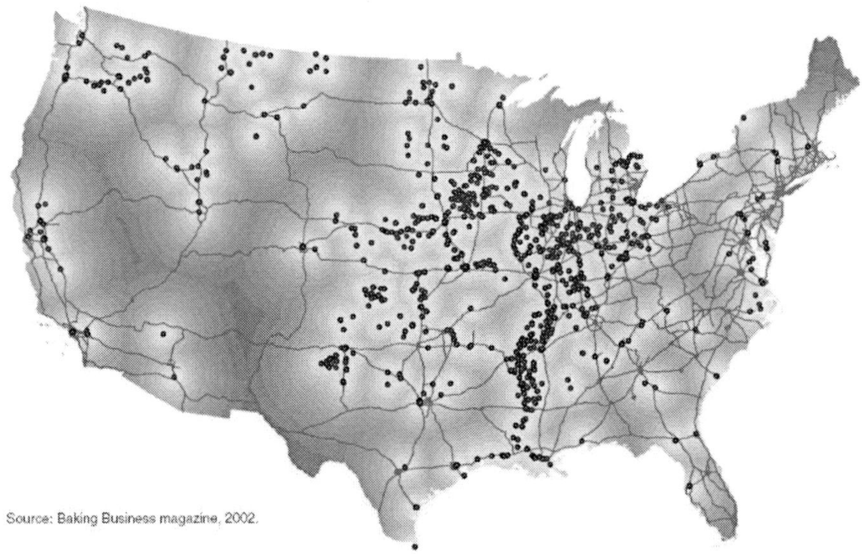

Figure 13. Spatial distrbution of grain elevators, 2002.

Correlation between Government Payments and Cropland Values Varies Regionally and by Payment Type

Government agricultural program payments may increase income from agricultural production. When they do so in a consistent way, the value of farmland is likely to increase due to expectations of a future income stream from government payments or reduced volatility in farm revenues (Gardner, 2003).[15] Although expectations of future government payments are difficult to observe, numerous studies have considered the degree to which payments under commodity programs are reflected in land values (see Moss and Schmitz, 2003, and references therein). Farmers have received commodity program payments for many decades, though the programs are subject to change and most must be reauthorized as part of the Farm Bill every 5 years. Over 2004-07, commodity payments averaged less than $25 per acre, but in some regions exceeded $100 per acre (USDA-ERS, 2011a).

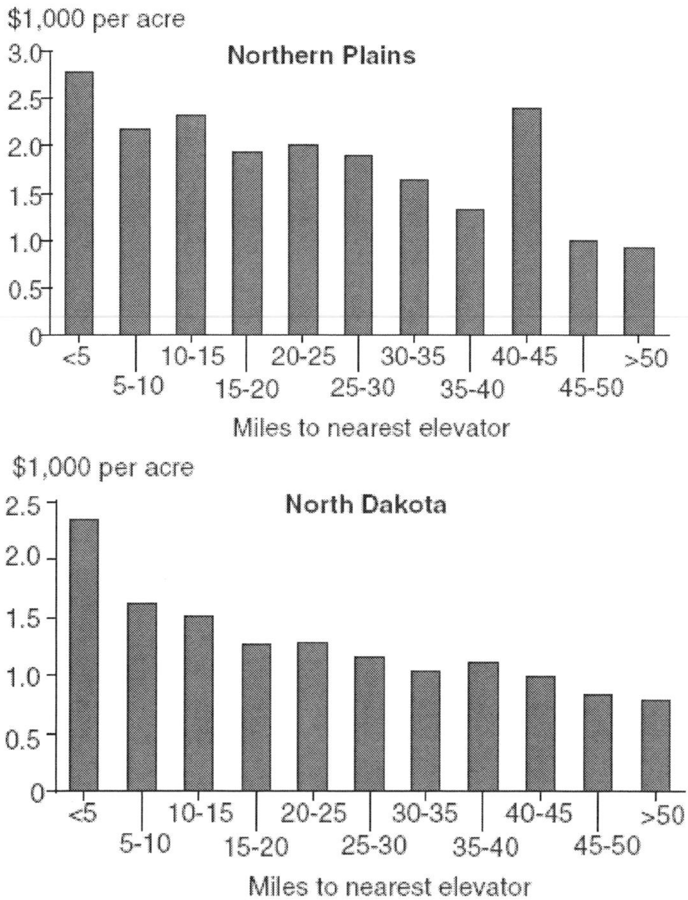

Sources: USDA, ERS analysis of National Agricultural Statistics Service June Area Survey data and grain elevator location data.

Figure 14. Cropland value per acre by distance to nearest elevator, Northern Plains and North Dakota, 2008.

The type of government payment or program can affect how it is reflected in farmland values. Payments that are certain might be accounted for differently than payments that are realized only under low prices, yields, or revenue. Production flexibility contract payments in effect for 1996- 2002—direct payments to producers with "base acreage" (i.e., historical rights to program benefits)—have been shown to increase cropland values (Goodwin et al., 2003; Roe et al., 2003). In contrast, programs that make payments based on low prices or yields, and thus support incomes, may not have as close a relationship with cropland values. Gardner (2003) indicates that a relationship may not be detectable because payments are non-existent when market prices are higher than pre-established target prices—i.e., at the time when land values increase the most. However, Goodwin et al. (2011) found that income support payments such as loan deficiency payments—which pay when market prices fall below the local loan rate— can add significantly to the value of land, due to reduced variability of farm earnings and more certain cash flow. Studies on the impact of government programs on cropland rents have commonly found that direct payments invariant to market conditions are more likely to be reflected in rental rates than those that vary based on market prices (Latruffe and LeMouel, 2009).

Two of the largest commodity programs enacted under the 2002 and 2008 Farm Acts (in terms of dollars spent) provide fixed direct payments and countercyclical payments. These programs make payments to farmers based on historical production (base acreage) of specific commodities such as corn, oilseeds, wheat, rice, and cotton, and payments are independent of farmers' cropping decisions in the current year.[16] Direct payments are fixed during the 5-year periods of the Farm Acts, while countercyclical payments are price-dependent and less predictable—payment is triggered by market prices falling below a pre-defined price, with the payment not known until after harvest. The impacts of different government payments on cropland values are difficult to predict based only on theoretical models, so empirical studies provide important insight. However, empirical evidence of the size of the impact has not been conclusive. For example, estimates of the impact of direct payments on land values range from negligible in some studies to 10-50 percent, depending on the region (Goodwin et al., 2003; Barnard et al., 1997).

Farms can also receive subsidized crop insurance premiums. Several Federal crop insurance plans—both for yield and revenue—are available to farmers. Almost all major commodities are covered by Federal crop insurance, which is administered by USDA's Risk Management Agency and delivered to farmers through private agents. Farmers pay 50 to 60 percent of the total crop insurance premium, and the areas for which crop insurance is available and the number of crops insured have been increasing over the past decade. Currently, over 270 million acres are insured, with subsidies of over $5 billion per year (USDA-ERS 2011a; 2011b).

Figure 15 shows average cropland values for land with different levels of direct and counter-cyclical program (DCP) payments[17] and crop insurance premium subsidies. Counties in the bottom 20th percentile of cropland acres are excluded to focus the analysis on more farm-dependent counties and to improve comparability of different payment levels. In 2007, crop insurance premium subsidies totaled $3.7 billion and DCP payments $6.2 billion. Although cropland values generally increase with insurance premium subsidies, the pattern is less consistent for DCP payments. Crop insurance premium subsidies may have a more consistent impact in recent years, because farmers have been receiving about $2 in indemnity payments for every $1 of premiums paid by farmers (Goodwin et al., 2011). This suggests that crop insurance premium subsidies and DCP payments might have different impacts on land values, but controlling for other factors affecting land values will be important.

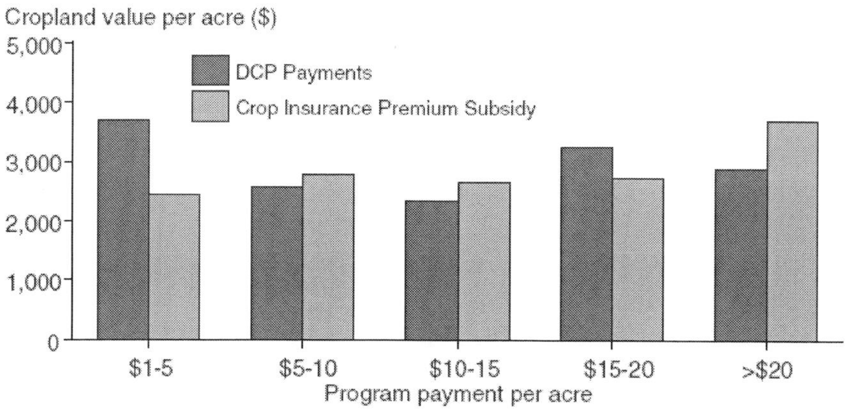

DCP = Direct and countercyclical program payments.

Source: USDA, ERS analysis of National Agricultural Statistics Service June Area Survey data,

Risk Management Agency and Farm Service Agency data, and Census of Agriculture data.

Figure 15. Cropland values by amount of selected government payments, 2007.

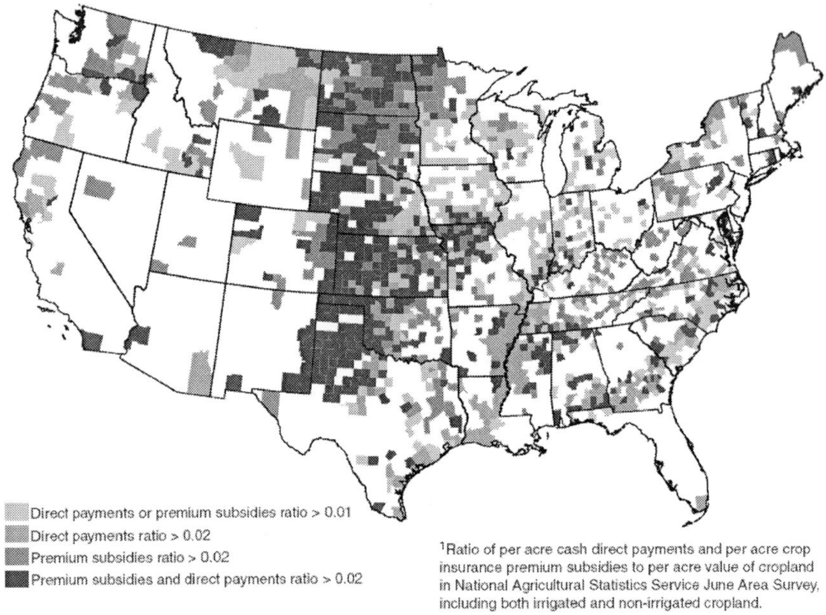

Source: USDA, ERS analysis of National Agricultural Statistics Service June Area Survey data, Risk Management Agency and Farm Service Agency data, and Census of Agriculture data.

Figure 16. Ratio of selected government payments to cropland value1, 2006-08.

There are considerable regional differences in the ratio of payments to cropland values (fig. 16).[18] Much of this regional diversity is likely based on differences in what crops are, and have historically been, grown. Cropland in the Delta and Southeast regions appears to have the largest level of DCP payments relative to cropland values. Cotton and rice production are high in these regions, especially the Delta (USDA-ERS, 2011c); DCP payments for these crops are the highest per acre, and the cotton base has most consistently received countercyclical payments. Some regions (the Northern and Southern Plains) have high levels of both premium subsidies and DCP payments relative to cropland values. The Corn Belt and Lake States have the lowest level of subsidies and DCP payments relative to land values. Areas without DCP payments but higher premium subsidies tend to be areas where fruits, vegetables, and other crops (sugar beets, other specialty crops) not part of the DCP historical program base are grown.

The need for additional empirical study of the impacts of government payments on farmland values—and how the impacts vary by payment type and by region—is particularly important given the current trend of high commodity prices, large government deficits, and debate over the future of agricultural payment programs. It is especially challenging to predict potential impacts on farmland values should programs be abruptly eliminated. Many studies have determined that agricultural payments raise farmland values. However, these studies use hedonic or other statistical regression approaches that capture only marginal changes (i.e., the change in per-acre values for a $1 change in payment) and do not take into account changes in commodity market prices, the emergence of replacement programs, and other significant consequences of eliminating agricultural programs. Nonetheless, our analyses suggest that abrupt changes in agricultural payment programs are likely to affect farmland values, with varying regional effects.

Nonfarm-Related Factors Affecting Farmland Values

Empirical studies of farmland values have found that in some areas of the country, returns from agricultural production are not the most significant contributor to farmland prices. Indeed, studies have estimated that well over half the variation in farmland prices is due to nonfarm factors in some States (Peterson, 1986; Plantinga et al., 2002).

Farm Real Estate Values Are Positively Correlated with Proximity to Urban Areas

The demand for land that can be developed for urban use is the most significant nonfarm factor affecting farmland values in areas that are more urbanized or are experiencing faster population growth (Livanis et al., 2006; Hardie et al., 2001). According to the classical model of urban form, land is allocated based on the needs of various economic agents who bid according to their desire to be close to markets or urban centers. The value of land, or location rent, declines as one moves away from urban centers.

While proximity to population centers and increased access to markets and customers could influence farmland values by increasing expected agricultural returns, development demand for residential, commercial, and industrial uses tends to increase farmland values even more. The U.S. population continues to grow and disperse, and all States are experiencing urban-related influences on farmland values to some degree. To acquire land for nonagricultural purposes, developers typically must bid higher than the agricultural production value of the land. As a result, even a limited number of conversions of farmland to urban uses can lead to generally higher farmland values in areas influenced by urban demand for land.

A typical approach to identify the amount of land subject to urban influence is to examine the location of land relative to the size and number of population concentrations. The Population-Interaction Zones for Agriculture (PIZA) measure developed by ERS is based on the concept of a gravity model (Shi et al.,1997) that accounts for both size of populations within a 50-mile radius, as well as distance between a land parcel and the populations (see Appendices A and B). Using the PIZA index, more than three-quarters of farm real estate (farm land and buildings) were estimated to be rural in 2008 (fig. 17). These are lands in tracts without any part of a town of 2,500 or more residents, and with the primary commuting pattern to sites within the tract. About one-fourth of farm real estate was estimated to be subject to some degree of urban influence.

Farm real estate values tend to decline the farther the land is situated from urban areas or population centers. Figure 18 shows the mean value of farm real estate, including all buildings and structures, at increasing distance from the borders of population centers. As one moves farther from a population center, the average farm real estate value per acre declines regardless of the size of the population center (50,000, 25,000, 10,000, or 5,000 residents). The gradient steeply declines at a distance less than 10 miles and then declines more gradually with increasing distance. Further, the mean values are higher for larger cities. For example, the average value for agricultural parcels within 5 miles of a city of at least 50,000 residents is approximately $16,801 per acre, whereas the average value the same distance from a town of at least 5,000 residents is $10,705. A major limitation of this definition of urban influence is that it provides little control for the relative size of population centers. For example, the definition of a city with at least 50,000 residents includes population centers such as Scranton, PA, with a population of just over 70,000 residents, and New York City, with a population of over 8 million residents. To mitigate this shortcoming and examine the robustness of the land value gradient, we evaluated two alternative measures of urban influence (see Appendix B). The additional analyses revealed very similar results.

Farm Real Estate Values Fall as Access to Road Transportation Declines

Economic theory suggests that access to transportation may play an important role in determining land values. Transportation infrastructure provides farmers with access to markets (urban centers, elevators, or other delivery points) and reduces input costs related to transportation. Figure 19 shows the average farm real estate value per acre relative to distance to major highways (interstates). Transportation effects may vary depending on urban influence, and the figure therefore divides the sample by PIZA classification (see Appendix B). The average value per acre exhibits a similar pattern across all PIZA classifications, declining as distance to interstate highways grows. The price decline is more pronounced at higher levels of population interaction.

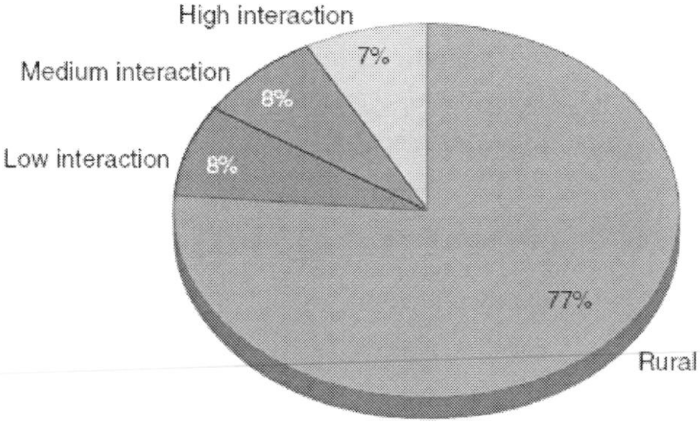

Note: Low, medium, and high interaction categories reflect the extent of interaction with population concentrations.

Source: USDA, ERS analysis of NASS June Area Survey data and PIZA data (see Appendix A). Agricultural real estate includes land and buildings.

Figure 17. Agricultural real estate subject to urban influence, 2008.

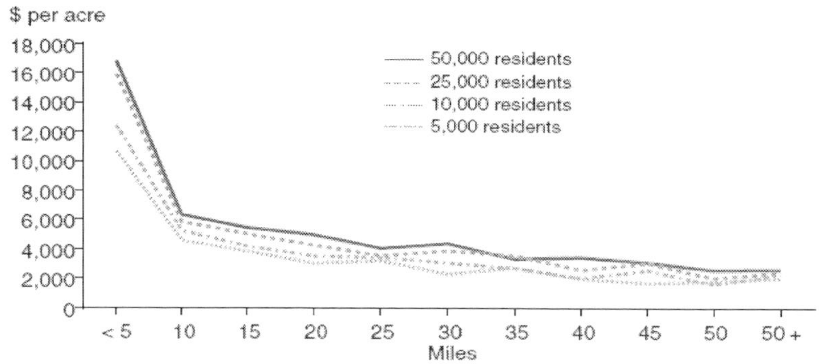

Sources: USDA, ERS analysis of National Agricultural Statistics Service June Area Survey data and Census of Population data. Agricultural real estate includes land and buildings.

Figure 18. Farm real estate values by distance to population centers, 2008.

Farmland Values May Not Be Strongly Linked with Rural Housing Markets

Farmland values are highly correlated with distance to urban areas. As a result, farmland values near urban areas are tied to the value of competing land use activities. Yet, because farmland provides a source of revenue through the production of agricultural goods and services, the pattern of aggregate farmland values and the values in developed uses (mainly residential uses) may differ by location according to agricultural production characteristics. One national analysis (Plantinga et al., 2002) using 1997 data found that the influence of development potential was highest in places with urban areas and little agricultural land. On average, the development value contributed about 10 percent to the total value of agricultural land, ranging from more than 30 percent in many Northeastern States and Florida (the contribution was more than 80 percent in New Jersey) to less than 5 percent in most of the Northern and Southern Plains States.

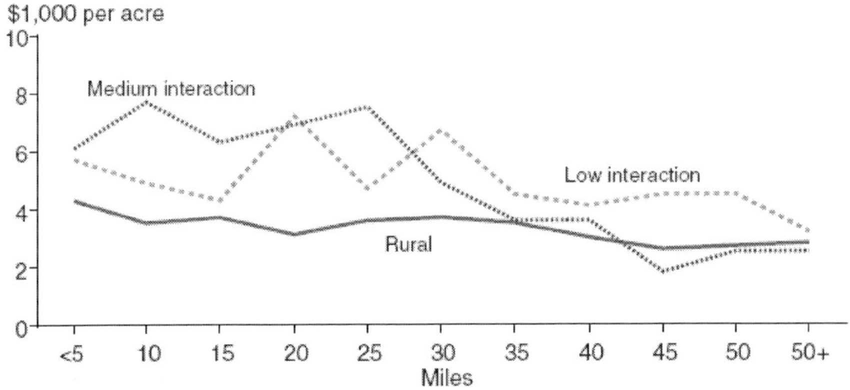

See appendix B for PIZA definitions.

Source: USDA, ERS analysis of National Agricultural Statistics Service June Area Survey data, PIZA data and interstate highway data (see Appendix A).

Figure 19. Farm real estate values by distance to nearest interstate highway by PIZA code, 2008.

The contribution of development value to farmland values raises questions about the extent to which farmland markets are susceptible to changes in residential markets. Over the last decade, the residential real estate market has undergone dramatic changes, with a housing boom that peaked in 2005 followed by a contraction. The contribution of residential investment to GDP dropped more than 20 percent from 2005 to 2007, with the contraction greatest in States where growth in residential home building was strong during the boom period (Siniavskaia, 2008). During 2001-04, California, Arizona, Oregon, Florida, Maryland, Connecticut, and Rhode Island all experienced increases of 20 percent or more in the Rural Housing Price Index (RHPI); in 2007-09, the RHPI in these States declined by at least 10 percent (fig. 20).[19] In the Northern and Southern Plains and Delta regions, the housing downturn dampened growth in the RHPI rather than reversing the gains over the 2001-04 period.

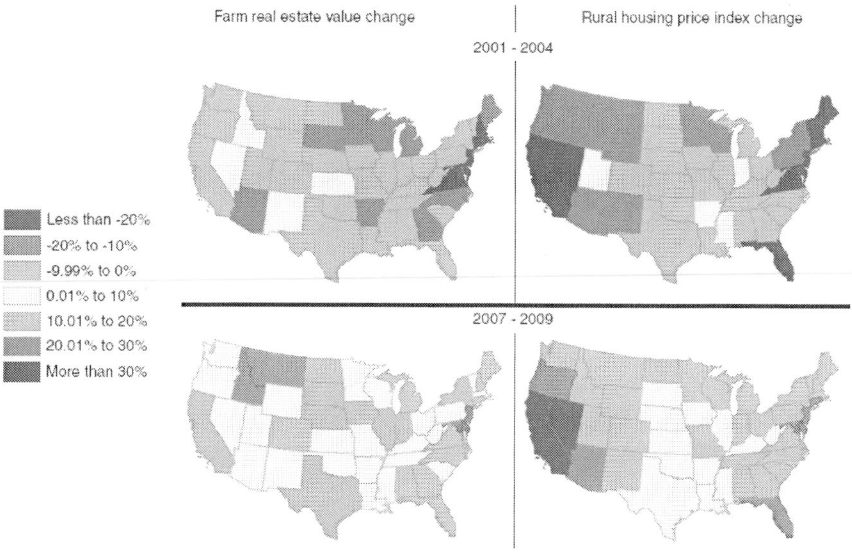

Sources: USDA, National Agricultural Statistics Service (2009a) and housing price index data (see footnote 19).

Figure 20. Changes in farm real estate values and rural housing values by State, 2001-04 and 2007-09.

A comparison of farm real estate values between the boom-and-bust periods reveals that farm real estate values increased in all States during 2001-04, and the gains were often as great or greater than gains in the rural housing market (fig. 20). From 2007 to 2009, farm real estate values declined in many States, but the reversals were generally more moderate than in the rural housing sector. In 19 States, notably California, Oregon, Washington, and Nevada, farm real estate values actually increased in 2007-09 while the RHPI declined. These trends suggest that strong gains in farm earnings and declining interest rates during 2007-09 helped the farmland market withstand the significant downturn in housing markets.

The relationship between farmland and rural housing markets is more evident when examining trends for particular States. The selected States (California, Iowa, and Massachusetts) represent three distinct areas in terms of agricultural production and residential housing. Average farm real estate value per acre and were obtained from the National Agricultural Statistics Service (USDA/NASS 2009a). Rural housing price index values were obtained from the Federal Housing Finance Agency (see footnote 19).

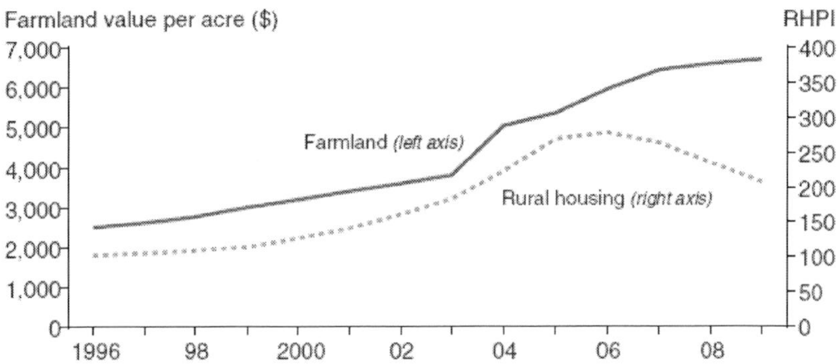

Source: USDA, ERS analysis of National Agricultural Statistics Service 2009a and housing price index data (see footnote 19).

Figure 21. Farm real estate values and rural housing price index (RHPI) in California, 1996-2009.

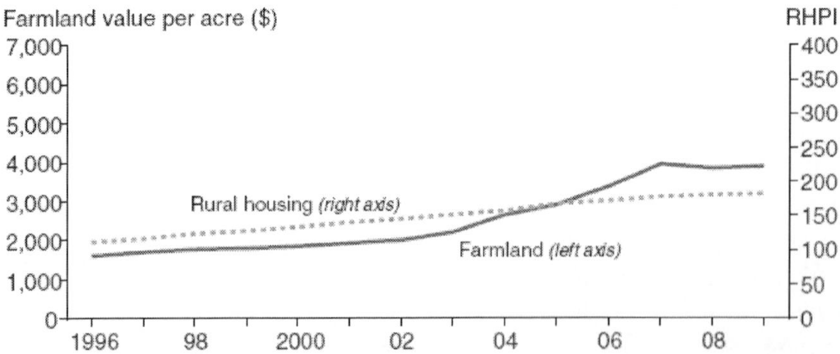

Source: USDA, ERS analysis of National Agricultural Statistics Service 2009a and housing price index data (see footnote 19).

Figure 22. Farm real estate values and rural housing price index (RHPI) in Iowa, 1996-2009.

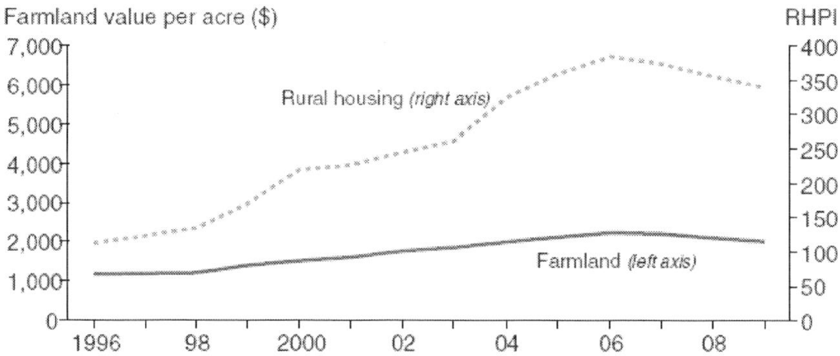

Source: USDA, ERS analysis of National Agricultural Statistics Service 2009a and housing price index data (see footnote 19).

Figure 23. Farm real estate values and rural housing price index (RHPI) in Massachusetts, 1996-2009.

California's real estate sector in 1996-2009 was characterized by rapid price appreciation followed by a downturn during the housing price bubble (fig. 21). The period of rapid price appreciation can also be observed in farm real estate values. From 1996 to -2007, both price series exhibited a similar average annual appreciation rate: 6 percent for rural housing and 8 percent for farm real estate. Farm real estate values continue to increase, albeit more slowly after the housing bubble burst in the late 2000s.

The price trends are more tempered in Iowa, where rural housing prices exhibited a moderate increase throughout 1996-2007, with an average annual increase of 4 percent (fig. 22). At the same time, farm real estate prices increased 7 percent per year, with gains particularly strong in 2004-07. Between 2007 and 2009, both price series leveled off.

As in California, rural housing prices in Massachusetts exhibited a boom-and-bust pattern in 1996-2009 (fig. 23). However, changes in farm real estate prices were less pronounced, increasing through 2006 and declining through 2009. Although the downturn is similar in magnitude for both housing and farm real estate values, the price appreciation was much greater for residential real estate. Rural housing prices appreciated 9 percent per year in 1996- 2009, versus 4 percent for farm real estate.

Natural Amenities Are Positively Related to Farm Real Estate Values

A number of studies suggest that like other land values, farmland values are influenced by the amount and nature of natural amenities, such as scenic views and varied terrain that attract people to an area (Pope, 1985). ERS' analysis of population changes suggests that high-amenity counties next to metropolitan areas are the most desirable places to live (USDA-ERS, 2009), so high-amenity areas could increase the development value of the land.

The Natural Amenities Scale published by the Economic Research Service offers a quantitative measure of landscape amenities at the county level for the contiguous United States (see Appendix A). The scale is constructed based on the notion that positive natural amenities include mild climate, varied topography, and proximity to surface water, including ponds, lakes, and shoreline. The scale is constructed using information on land surface topography, winter temperatures and sunlight, summer temperatures and humidity, and water cover. The Natural Amenities Scale is categorized by deviations from mean amenity levels, ranging from 1 (low amenity) to 7 (high amenity). The scale has been used in previous research, including a study of the effects of natural amenities on rural population change (McGranahan, 1999).

High levels of natural amenities have a large positive effect on farm real estate values, with a mean value per acre of $58,732 at the highest level of the index value (fig. 24). At the lowest levels of natural amenities, mean values are $3,351 per acre. The impact of natural amenities falls greatly from the highest classification (7) but remains relatively constant at lower levels (1-5).

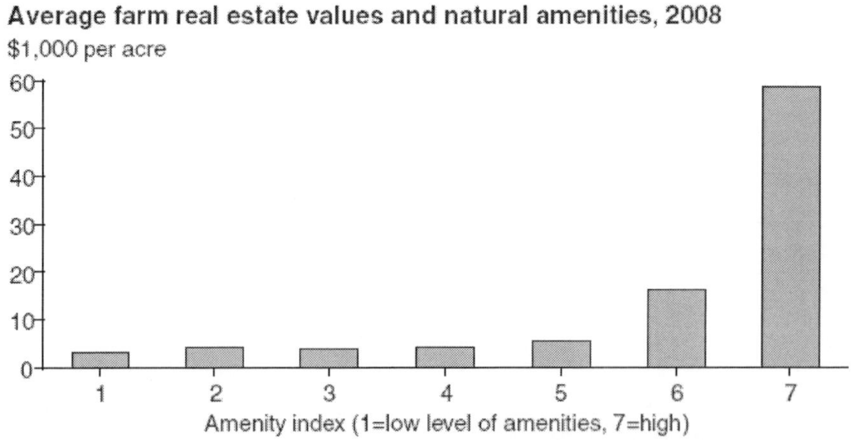

Source: USDA, ERS analysis of National Agricultural Statistics Service June Area Survey data and ERS Natural Amenity Scale data.

Figure 24. Average farm real estate values and natural amenities, 2008.

While natural amenities arising from topography and climate conditions are correlated with land values, other amenities arise from the farmland itself and tend not to be reflected in the land's value. These "rural" amenities include the habitat provision, groundwater recharge, and open-space benefits provided by having land in an agricultural use as opposed to a more intensive use (Hellerstein et al., 2002; Irwin et al., 2003). Though the landowner cannot typically capture the value associated with rural amenities through selling land on the market, evidence suggests that people do value these public goods and they are a primary reason for public support for farmland protection policies (e.g., Kline and Wichelns, 1996). Demand for protecting rural amenities is highest in suburban and urban fringe areas where the loss of agricultural lands to urban uses is highest (Nickerson and Hellerstein, 2003; Kline and Wichelns, 1994).

While the rural amenity value of land is not reflected in farmland values, some policies to protect rural amenities can affect farmland values. State and local governments in every State have enacted measures to help retain farmland in an agricultural use and to prevent the loss of rural amenities (Hellerstein et al., 2002). Preferential assessment programs, in which landowners get property tax relief for keeping the land in agriculture, generally have little effect on farmland values—likely because many such programs contain penalties wherein a portion of the property tax savings must be paid upon conversion to a developed use (e.g., Parks and Quimio, 1996; Wunderlich, 1997). Regulatory policies such as zoning often specify minimum lot sizes, and many zoning laws restrict the ability of landowners to convert farmland to urban uses. However, studies show that such policies reduce land values for the restricted farm parcels when development pressure is present (Gottlieb and Adelaja, 2009; Beaton, 1991).

In some cases, impacts of farmland protection vary by location. Wisconsin's exclusive agricultural zoning policy, which helps contain urban development, was found to decrease per-acre farmland prices for smaller agriculturally zoned parcels located closer to cities but increase prices for larger parcels far from cities (Henneberry and Barrows, 1990). Purchase of development rights programs, in which landowners voluntarily sell the rights to develop the land in exchange for a payment, result in permanent (or very long term, 25-30 year) restrictions on nonfarm development. In theory, once the development rights have been sold, the market value of the preserved land will reflect only its value in a farming use, which in urbanizing areas can be significantly lower than its residential market value. However, several studies have found little evidence that such development restrictions significantly lowered preserved farmland prices (Nickerson and Lynch, 2001; Lynch et al., 2007; Anderson and Weinhold, 2008). The authors speculated that landowners who farm as a recreational pursuit are outbidding "traditional" farmers for the preserved land.[20]

FARMLAND OWNERSHIP TRENDS

The ownership of agricultural land can have far-reaching implications on the food and fiber system. The most recent estimate of the number of farmland owners was in 1999, when USDA-NASS estimated that there were more than 2.2 million agricultural landlords.[21] Ownership arrangements can affect farmland values through their impacts on the decisionmaking behavior of farm operators, including production decisions, adoption of technologies, and conservation practices that can enhance the productivity of the land. For example, operators who are full tenants and own none of the land they operate are more likely to specialize in crops (versus livestock) than operators who own the land they operate (Hoppe et al., 2007). While there has been longstanding interest in the characteristics of farmland owners and the extent to which they are actively engaged in farming, this interest has increased in recent years with the increase in farmland prices and uncertainty over whether farmers or nonfarmers are reaping the benefits of farm program payments.[22]

The amount of land in farms that is rented by farm operators has fluctuated over time, ranging between 34 percent and 43 percent of farmland over 1964-2007 (fig. 25). The rising share of rented land from 1982 (38 percent) to 1992 (43 percent) was partly due to the farm crisis in the early 1980s that forced some farms out of business and the owners to rent their land. However, since 1997 the amount of acreage rented has remained below 40 percent. The distribution of rented farmland varies across regions, with the highest proportions in Illinois, Iowa, Kansas, and the Mississippi Delta. Except for increases in a few areas (the Mississippi Delta region, southern Illinois, eastern North Dakota, and the Central Valley in California), the geographic concentration of counties with at least half of farmland rented or leased in 2007 was similar to 1950 (fig. 26). With such a large amount of farmland rented, landlords clearly play a significant role in U.S. agriculture, particularly in some regions.

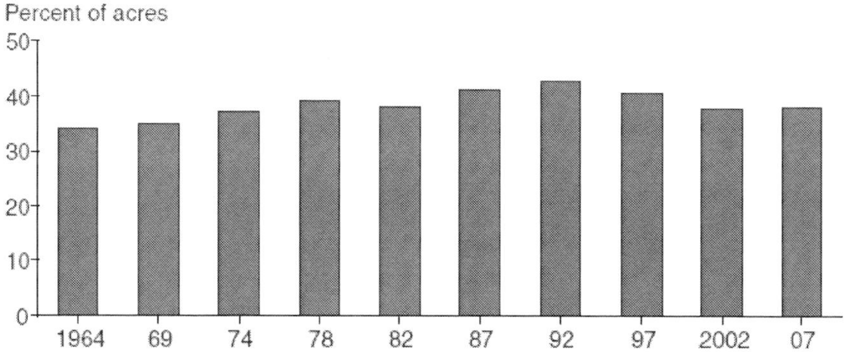

Source: Census of Agriculture, multiple years.

Figure 25. Share of farmland acres rented or leased, 1964-2007.

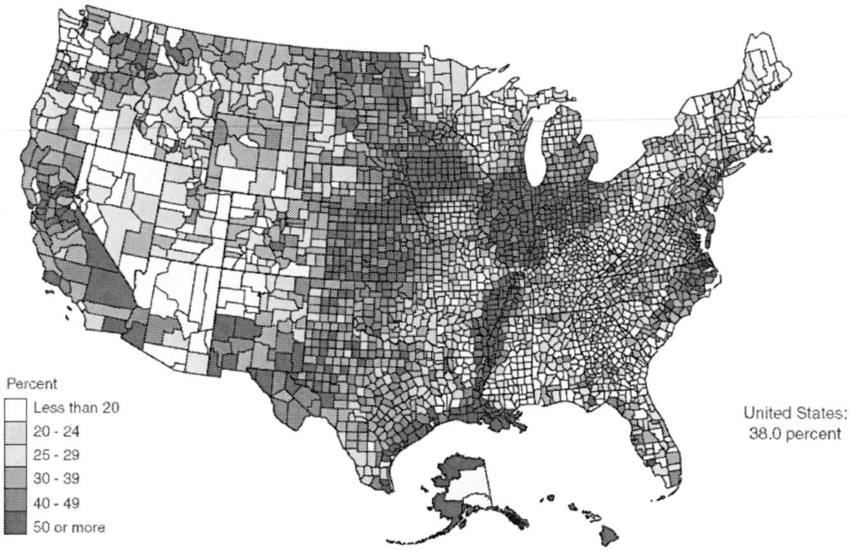

Source: USDA, National Agricultural Statistics Service.

Figure 26. Percent of land in farms rented or leased, 2007.

Farmland owners can be grouped into two categories: owner-operators (those owning all the land they operate; this includes farmers who operate some land and rent out additional land to other farmers) and non-operating land owners (NOLs). The proportion of land in farms owned by NOLs declined from 37 percent in 1992 to 29 percent in 2007 (USDA-NASS, 2009). The amount of acres held by NOLs differs across regions (fig. 27). In 2007, in three of the top four regions in terms of land in agriculture—the Corn Belt,

Northern Plains, and Southern Plains—NOLs owned more than 30 percent of the land. NOLs also owned 30 percent or more of farmland in the Delta and Pacific regions. Many reasons contribute to this pattern, including laws in some States that limit farmland ownership by nonfarmers or by foreign owners.

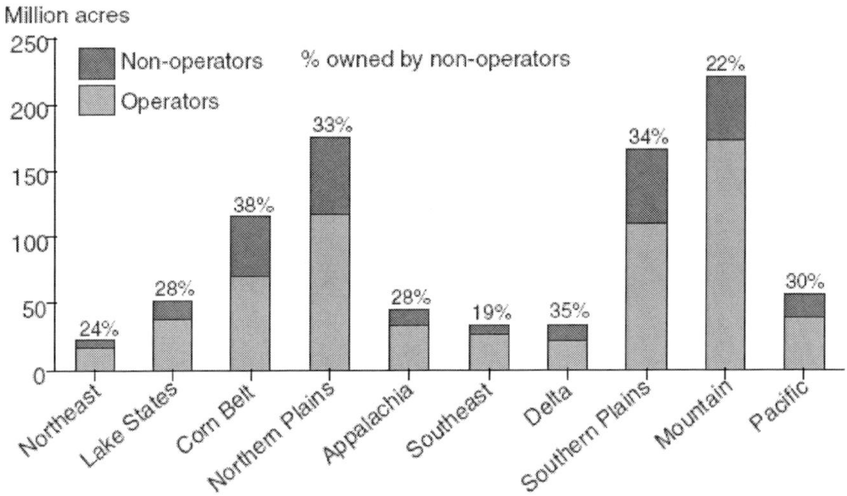

Source: Census of Agriculture (2007).

Figure 27. Farmland owned by operators and non-operator landlords by region, 2007.

Source: USDA, National Agricultural Statistics Service Agriculture Economics and Land Ownership Survey (AELOS), 1999.

Figure 28. Number of non-operating owners and the acres they own, 1999.

The characteristics of farm operators have been well documented (Hoppe et al., 2007), but much less is known about non-operator landlords. Data from a survey of private farmland owners in 1999 (the most recent available) revealed that most land owned by NOLs was concentrated in the hands of relatively few people who owned large amounts of land. In 1999, 4 percent of NOLs owned at least 1,000 acres of farmland, and collectively they controlled 44 percent of all acreage owned by NOLs (fig. 28). However, about 68 percent of NOLs owned less than 180 acres each; these landlords owned 18 percent of the land owned by NOLs. Using a combination of the two most recent surveys of private farmland owners (1988 and 1999) and 2007 Census of Agriculture data indicates whether characteristics of farmland owners are changing over time. Age of farmland owners is of particular interest because land turnover is higher among older owners. A greater proportion of non-operators (than operators) are older than 70: 47 and 42 percent of NOLs were older than 70 in 1988 and 1999, respectively, while only about 20 percent of owner-operators were older than 70 over the past three decades. Far fewer NOLs live on the farm, compared with owner-operators (29 versus 82 percent in 1999), though another 29 percent of NOLs live within 5 minutes of the farm they own (Barnard et al., 2001). It also appears NOLs were less likely to be enrolled in the Conservation Reserve Program (CRP) or the Wetlands Reserve Program (WRP), at least in the 1980s and 1990s.[23] The proportion of owner-operators participating in the CRP or WRP increased in 1999-2007 from 10 percent to 16 percent; national data are not available to determine whether enrollment patterns among non-operators changed after 1999. This is consistent with research that found that tenants were less likely than owner-operators to adopt conservation practices that provide benefits over the longer term (Soule et al., 2000). In 1999, non-operators also tended to own farm real estate that was about $100 per acre less valuable, on average, than land owned by owner-operators.

Many Non-Operating Landlords Live in Towns and Urban Areas

The Agricultural Economics and Land Ownership Survey (AELOS) indicates that in 1999 (the last update), 42 percent of all farm landlords lived in a city, town, or urban area.[24] Some farm program payments are paid directly to landlords (see footnote 16), and when a large percentage of farmland owners may live in urban areas, it raises questions about whether—in addition to supporting farm incomes—farm program payments benefit the rural economy as intended (Drabenstott, 2005). However, ERS analysis of 2004 data on government payments made in Federal commodity and conservation programs revealed that over 90 percent of payments in 2004 were mailed to rural addresses outside of census-defined urban areas and clusters.

Table 2. Characteristics of the average farmer and non-operating landowner

	1988		1999		2007
Characteristi	Non-Operator	Owner-Operator	Non-Operator	Owner-Operator	Owner-Operator
	Percent				
70 years or older	47.7	21.7	42.1	17.6	19.1
Live on farm	36.3	78.6	29.0	82.2	76.8
Enrolled in the Conservation or					
Wetlands Reserve Program	8.1	11.1	8.0	10.4	15.7

Sources: USDA, National Agriculture Statistics Service Agricultural Economics and Land Ownership Survey (1988 and 1999) and Census of Agriculture (2007).

Relatively Little U.S. Farmland Is Foreign Owned

High levels of foreign demand for farmland could put upward pressure on farmland prices. Despite recent increases in foreign ownership of forest land, annual data compiled by USDA's Farm Service Agency reveals that, as of February 2009, only 1.7 percent of privately owned land in farms or forest, or 22.8 million acres, was owned by foreigners (fig. 29). The majority of foreign-owned land is owned by people or firms from two countries: Canada (34 percent) and the Netherlands (17 percent). Most foreign ownership is concentrated in Maine: 15.7 percent of Maine's privately owned farm and forest land, or 2.82 million acres, is foreign owned. Of this amount, 2.77 million acres are forest land. Other States with relatively high levels of foreign ownership of farm and forest land include Hawaii (8.8 percent), Washington (7.6 percent), Nevada (5.2 percent), and Alabama (5.1 percent). Foreign interests own 2.57 million acres in Texas.

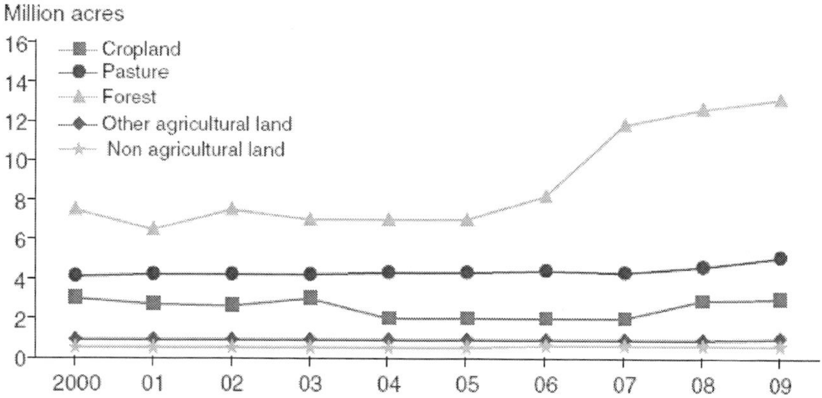

Source: Blevins et al., 2010.

Figure 29. Trends in foreign holdings of U.S. farmland, 1999-2009.

CONCLUSION

Farmland represents a significant investment and a large portion of the wealth of many U.S. farmers. With farmland real estate values growing largely steadily since the farm financial crisis in the mid-1980s, interest in the driving forces behind farmland values has grown, especially amid the residential housing downturn. Since 2005, increases in crop prices and cash rents, coupled with low interest rates and improved agricultural credit conditions, are giving rise to double-digit increases in farmland values in some regions.

Historical relationships between cropland and pastureland values have changed in some regions. Cropland values have typically exceeded pasture values and have been at least twice as large as pasture values in the Midwest and West in the last decade. But the premium has shrunk recently, and in two regions—the Southeast and Delta—average cropland values have remained below average pasture values since 2004-05. Studies suggest that nonfarm factors could be contributing to the relative changes in crop and pasture values. Pastureland can provide a stream of recreational income from hunting leases, and in the Southeast, pastureland tends to be closer to urban areas and development demand.

How has farmland been performing as an investment? Trends in farm incomes, cash rents and interest rates suggest farmland values were supported by farm earnings in 2009 and 2010. This suggests that in very recent years (when measured at the national level), farmland values are supported by fundamental farm factors. Yet, several macroeconomic measures indicate that over a longer horizon, farmland values are becoming less correlated with farm-related factors once thought to support those values. Declining rent-tovalue ratios indicate cash rents are increasingly smaller relative to farmland values, and the ratio is smallest for cropland close to urban areas. Also, the affordability of farmland has varied over time. While in 2009-2010 average income from farming has been more than sufficient to service farm real estate debt, during 2005-08 and during 1978-1985, this was not the case. A lack of correlation with net farm incomes, declining rent-to-value ratios, and low levels of affordability all suggest that nonagricultural factors are increasingly important in determining farmland values.

Farm program payments are under increased scrutiny amid a historic debate over the size of the Federal deficit. Our analyses suggest that changes to some of the largest farm programs would likely have impacts that vary across regions. Direct and countercyclical (DCP) payments are a larger percentage of cropland values in the Delta and Southeast regions than elsewhere. While the extent to which farm program payments increase farmland values is uncertain, studies indicate at least some capitalization of farm program payments. Large reductions in DCP payments might have the greatest potential to reduce land values in the Delta and Southeast. The Corn Belt and Lake States regions generally have the lowest level of subsidies (through both the DCP program and crop insurance) relative to land values. In the Appalachia and Northeast regions, some areas have high levels of insurance premium subsidies relative to land values, but some areas also have low levels of DCP payments relative to land values.

How does farmland compare to alternative investments? While farmland has historically been a stable investment, it has been outperformed by the S&P 500 (and gold, when investments have been held for nearly 40 years). Yet in very recent years, farmland has become more attractive as an investment, as reflected in surveys that indicate active investor interest in farmland, at least in some regions: nearly a quarter of farmland purchases in Iowa were made by investors in 2009. However, uncertainty about the renewal of major farm commodity programs and future interest rates could reduce the attractiveness of farmland as an investment relative to other alternatives.

While macroeconomic conditions affect farmland values, farmland values are also driven by parcel-specific factors that collectively determine the land's potential returns from being in an agricultural use. The importance of using spatially disaggregated data, though complex and time consuming, is particularly evident when examining correlations between common proxies for net returns to agricultural production. Our analyses reveal that the positive relationships between better soils and cropland values became evident only when we controlled for proximity to large population concentrations (we retained land value observations only in the most rural areas). Parcel-specific effects also vary substantially across regions, with opposite effects in some cases (e.g., soil quality was negatively correlated with cropland values in the Appalachian region but positively correlated elsewhere). Cropland values generally increase with proximity to points of delivery (grain elevators, highways) in rural areas.

Significant volatility in residential real estate markets over the last decade has raised questions about the extent to which competing land markets affect farmland values. A comparison of rural housing and farmland value trends in California, Iowa, and Massachusetts—three States that are distinct in terms of location, agricultural production and residential housing—reveals that the boom-and-bust cycle in the housing market in the last decade had no, or relatively modest, negative impacts on farmland markets. Of the three States, only farmland values in Massachusetts declined between 2006 and 2009, coincident with rapid losses in residential housing values. These comparisons suggest that farmland markets may be fairly independent of residential housing markets in some places but not in others. However, the moderating impacts of historically low interest rates and high crop prices almost certainly mitigated the downturn in value that farmland markets could have otherwise experienced.

Coupled with questions about the likely stability of farmland values are questions about whether active farmers benefited from the recent increases in those values. With about 40 percent of farmland being rented, and with more than 30 percent being rented from non-operating landowners in several regions (largely the Midwest and Pacific), it is clear that non-operating landowners play a significant role in U.S. agriculture. Not only do non-operators reap some of the benefits (and bear some of the risks) of changes in farmland values, land owned by non-operators is managed somewhat differently—for example, such lands may be less likely to be enrolled in conservation programs.

The trend analyses in this report provide an important foundation in understanding the factors driving farmland values and ownership. Nonetheless, the correlations of values and tenure status with macroeconomic and parcel-specific factors do not necessarily identify causal relationships. Many of the factors affecting land values and ownership are changing simultaneously, and it is difficult to know from trend analysis alone which or to what extent particular factors are contributing to changes. Additional research that statistically analyzes these complex interactions can help identify which of the myriad factors have the most significant impacts on farmland values and ownership.

REFERENCES

Alston, J. 1986. "An Analysis of Growth of U.S. Farmland Prices: 1963-1982." *American Journal of Agricultural Economics* 68:1-9.

Anderson, Kathryn, and Diana Weinhold. 2008. "Valuing Future Development Rights: The Costs of Conservation Easements," *Ecological Economics*, Dec., v. 68 (1-2):437-46.

Barnard, C., R. Nehring, J. Ryan, R. Collender, and B. Quinby. 2001. "Higher Cropland Value from Farm Program Payments: Who Gains?" *Agricultural Outlook*, Nov., J. McLaughlin (ed.), U.S. Department of Agriculture, Economic Research Service.

Barnard, C., G. Whittaker, D. Westenbarger, and M. Ahearn. 1997. "Evidence of capitalization of direct government payments into U.S. cropland values," *American Journal of Agricultural Economics* 79.

Barnett, Barry J. 2010. "The U.S. Farm Financial Crisis of the 1980s," *Agricultural History* 74(2):366-380.

Barry, Peter J. 1980. "Capital Asset Pricing and Farm Real Estate," *American Journal of Agricultural Economics* 62:547-53.

Bastian, Chris T., Donald M. McLeod, Matthew J. Germino, William A. Reiners, and Benedict J. Blasko. 2002. "Environmental Amenities and Agricultural Land Values: A Hedonic Model Using Geographic Information Systems Data," *Ecological Economics* Nov., 40: 337-349.

Beaton, W.P. 1991. "The Impact of Regional Land-Use Controls on Property Values: The Case of the New Jersey Pinelands," *Land Economics* 67(May):172-94.

Beattie, G., and R. Ransom. 1979. "Use Value Assessment: Its Causes, Its Characteristics, Its Effects," Lincoln Institute of Land Policy, Cambridge, MA.

Blevins, Patricia, Lesa Johnson, and Dolores Smith. 2010. "Foreign Holdings of U.S. Agricultural Land through February 28, 2009." USDA Farm Service Agency, http://www.fsa.usda.gov/Internet/FSA_File/ afida_2010.pdf

Briggeman, Brian, and Maria Akers. 2010. "Farmland Values Climb and Credit Conditions Improve." Survey of 10th District Agricultural Credit Conditions. Federal Reserve Bank of Kansas City, 3rd quarter. http://www.kansascityfed.org/publicat/research/indicatorsdata/agcredit/AGCR3Q10.pdf

Burt, O.R. 1986. "Econometric Modeling of the Capitalization Formula for Farmland Prices," *American Journal of Agricultural Economics* 68(1):10-26.

Calhoun, C.A. 1996. "OFHEO House Price Indexes: HPI Technical Description," Office of Federal Housing Enterprise Oversight. http://www. fhfa.gov/webfiles/896/hpi_tech.pdf

Chambers, R. 1995. "The incidence of agricultural policies," *Journal of Public Economics* 57, 317-335.

Chicoine, D.L., and J.T. Scott. 1983. "Agricultural use-valuation using farm level data," *Property Tax Journal*, 2, 1-12.

Chavas, Jean-Paul, and C. Richard Shumway. 1982. "A Pooled Time-Series Cross-Section Analysis of Land Prices," *Western Journal of Agricultural Economics* 7(1):31-41.

Chavas, J.P., and A. Thomas. 1999. "A Dynamic Analysis of Land Prices," *American Journal of Agricultural Economics* 81(4):772-884.

Chicoine, David L. 1981. "Farmland Values at the Urban Fringe: An Analysis of Sale Prices," *Land Economics* 57 (Aug.):352-62.

Clark, J.S., M. Fulton, and J.T.Scott, Jr. 1993. "The Inconsistency of Land Values, Land Rents, and Capitalization Formulas," *American Journal of Agricultural Economics* 75(1):147-55.

Conklin, H.E., and W. G. Lesher. 1977. "Farm value assessment as a means for reducing premature and excessive agricultural disinvestment in urban fringes," *American Journal of Agricultural Economics* 59:755-759.

Duncan, Marvin. 2008. "Financing Agriculture in the 1980s," *Economic Review*, Federal Reserve Bank of Kansas City, March.

Drabenstott, M. 2005. "Do Farm Payments Promote Rural Economic Growth?" Center for the Study of Rural America, Federal Reserve Bank of Kansas City.

Dobbins, C., and K. Cook. 2007. *Indiana Farmland Values & Cash Rents Jump Upward*, Purdue Agricultural Economics Report 1-8, Aug.

Duffy, M. 2009. Farmland Value Survey, *Ag Decision Maker* File C2-70. Iowa State University Extension, Dec.

Falk, B. 1991. "Formally Testing the Present Value Model of Farmland Prices," *American Journal of Agricultural Economics* 73(1):1-10.

Featherstone, A.M., and T.G. Baker. 1987. "An Examination of Farm Sector Real Asset Dynamics," *American Journal of Agricultural Economics* 69:532-46.

Flanders, A., F. White, and C. Escalante. 2004. "Equilibrium of Land Values from Agricultural and General Economic Factors for Cropland and Pasture Capitalization in Georgia," *Journal of Agribusiness* 22: 49-60.

Freeman, M. 1993. *The Measurement of Environmental and Resource Values: Theory and Methods*. Washington, DC: Resources for the Future.

Gardner, B. 2003. "U.S. Commodity Policies and Farmland Values," in *Government Policy and Farmland Markets: The Maintenance of Farmer Wealth*, Charles B. Moss and Andrew Schmitz (eds.), Iowa State Press, IA.

Goodwin, B., A. Mishra, and F. Ortalo-Magné. 2011. "The Buck Stops Where? The Distribution of Agricultural Subsidies." National Bureau of Economic Research Working Paper 16693, http://www.nber.org/papers/w16693

Goodwin, B.K., A.K. Mishra and F.N. Ortalo-Magné. 2003. "Explaining Regional Differences in the Capitalization of Policy Benefits into Agricultural Land Values," in *Government Policy and Farmland Markets: The Maintenance of Farmer Wealth*, Charles B. Moss and Andrew Schmitz (eds.), Iowa State Press, IA.

Gottlieb, P.D., A. Adelaja. 2009. "The Impact of Down-Zoning on Land Values: A Theoretical Approach," *Agricultural Finance Review* 69(2):206-27.

Guiling, P., B.W. Brorsen, and D. Doye. 2007. "How Much Influence Does Recreation Have on Agricultural Land Values?" Selected paper prepared for presentation at the Southern Agricultural Economics Association Annual Meeting, Mobile, AL, Feb. 4-7.

Hardie, Ian, T. Narayan and B. Gardner. 2001. "The Joint Influence of Agricultural and Nonfarm Factors on Real Estate Values: An Application to the Mid-Atlantic Region." *Amer. J. Agr. Econ.* 83(1) (February 2001): 120–132.

Hellerstein, D., C. Nickerson, J. Cooper, P. Feather, D. Gadsby, D. Mullarkey, A. Tegene, and C. Barnard. 2002. *Farmland Protection: The Role of Public Preferences for Rural Amenities.* USDA-ERS Agricultural Economic Report AER815, November. http://www.ers.usda.gov/publications/aer815/

Henderson, Jason. 2010. "Will the Rural Economy Rebound in 2010?" *Federal Reserve Bank of Kansas City Economic Review*, 95(1):95-119.

Henderson, Jason, and Sean Moore. 2006. "The Capitalization of Wildlife Recreation Income into Farmland Values," *Journal of Agricultural and Applied Economics* 38(3):597-610 (Dec.).

Henderson, Jason, and Brent A. Gloy. 2009. "The Impact of Ethanol Plants on Cropland Values in the Great Plains," *Agricultural Finance Review*, v. 69, issue 1, pp. 36-48.

Henderson, J., and M. Akers. 2009. "Will High Farmland Values Hold?" *The Main Street Economist*, 4:6, Federal Reserve Bank of Kansas City.

Henderson, J., and N. Novack. 2005. "A Resurgent Rural Economy Spurs Farmland Values," *Federal Reserve Bank of Kansas City Economic Review*, 1st quarter, v. 90 (1):59-82.

Henneberry, D.M., and R.L. Barrows. 1990. "Capitalization of Exclusive Agricultural Zoning into Farmland Prices," *Land Economics* 66 (Aug.): 249-58.

Hoppe, R., P. Korb, E.J. O'Donoghue, and D.E. Banker. 2007. *Structure and Finances of U.S. Farms, 2007 Edition*. EIB- 24, U.S. Department of Agriculture, Economic Research Service. June.

Hoppe, R. 2006. "Land Ownership and Farm Structure," in *Agricultural Resources and Environmental Indicators*, Noel Gollehon and Keith Weibe (eds), EIB-16, U.S. Department of Agriculture, Economic Research Service.

Huang, Haixiao, Gay Y. Miller, Bruce J. Sherrick, and Miguel I. Gomez. 2006. "Factors Influencing Illinois Farmland Values," *American Journal of Agricultural Economics*, May, v. 88, issue 2, pp. 458-70.

Irwin, E., C. Nickerson, and L. Libby. 2003. "What Are Farmland Amenities Worth?" *Choices*, 3rd quarter.

Irwin, S., L. Forster, and B. Sherrick. 1988. "Return to Farm Real Estate Revisited." *American Journal of Agricultural Economics* 70:580-87.

Kline, J., and D. Wichelns, 1996. "Public preferences regarding the goals of farmland preservation programs," *Land Economics*, 72: 538-549.

Kline, J., and D. Wichelns. 1994. "Using referendum data to characterize public support for purchasing development rights to farmland," *Land Economics*, 70(2): 223-33.

Latruffe, L., and C. Le Mouel. 2009. "Capitalization of Government Support In Agricultural Land Prices: What Do We Know?" *Journal of Economic Surveys* 23: 659-691.

Libby, L.W., and E.G. Irwin. 2003. "Rural Amenities and Farmland Values," in *Government Policy and Farmland Markets: The Maintenance of*

Farmer Wealth, Charles B. Moss and Andrew Schmitz (eds.), Iowa State Press, IA.

Livanis, Grigorios, Charles B. Moss, Vincent E. Breneman, and Richard F. Nehring. 2006. "Urban Sprawl and Farmland Prices," *American Journal of Agricultural Economics*, Nov. 88(4): 915-29.

Luttrell, C.B. 1979. "The 'Danger' From Foreign Ownership of U.S. Farmland," *Federal Reserve Bank of St. Louis Review*, Jan.:2-9.

Lynch, L., W. Gray, and J. Geoghegan. 2007. "Are farmland preservation program easement restrictions capitalized into farmland prices? What can a propensity score matching analysis tell us?" *Review of Agricultural Economics* 29 (3): 502-509.

McGranahan, D. 1999. *Natural Amenities Drive Rural Population Change.* AER-781, U.S. Department of Agriculture, Economic Research Service, http://www.ers.usda.gov/publications/aer781/aer781.pdf

McNew, K., and D. Griffith. 2005. "Measuring the Impact of Ethanol Plants on Local Grain Prices," *Review of Agricultural Economics*, Summer, v. 27, issue 2, pp. 164-80.

Melichar, E. 1984. "A Financial Perspective on Agriculture." *Federal Reserve Bulletin* 70:1-13.

Monke, J. 2006. "Farm Commodity Programs: Direct Payments, Counter-Cyclical Payments, and Marketing Loans." CRS Report for Congress, Order Code: RL33271.

Moss, Charles B., and Andrew Schmitz (eds.). 2003. *Government Policy and Farmland Markets: The Maintenance of Farmer Wealth.* Iowa State Press, IA.

Moss C.B., and A.L. Katchova. 2005. "Farmland Valuation and Asset Performance," *Agricultural Finance Review* 65(2):119–30.

Nickerson, C.J., and D. Hellerstein. 2003. "Protecting rural amenities through farmland preservation programs," *Agricultural and Resource Economics Review*, 32(1): 129-144.

Nickerson, C.J., and L. Lynch. 2001. "The Effect of Farmland Preservation Programs on Farmland Prices," *American Journal of Agricultural Economics*, 83(2)(May):341-351.

Oppendahl, D.B. 2010. "Farmland Values and Credit Conditions," *The Agricultural Newsletter from the Federal Reserve Bank of Chicago*, n. 1950, Nov. http://www.chicagofed.org/digital_assets/publications/agletter/2010_2014/november_2010.pdf

Palmquist, R., and L.E. Danielson. 1989. "A Hedonic Study of the Effects of Erosion Control and Drainage on Farmland Values," *American Journal of Agricultural Economics* 71(1) (Feb):55-62.

Parks, P.J., and W.R.H. Quimio. 1996. "Preserving Agricultural Land with Farmland Assessment: New Jersey as a Case Study," *Agricultural and Resource Economics Review* April, 25(1): 22-27.

Peterson, W. 1986. "Land Quality and Prices," *American Journal of Agricultural Economics,* Nov. 68(4): 812-819.

Plantinga, A., Ruben N. Lubowski, and Robert N. Stavins. 2002. "The effects of potential land development on agricultural land prices," *Journal of Urban Economics* 52:561-581.

Pope, C.A. III. 1985. "Agricultural Productive and Consumptive Use Components of Rural Land Values in Texas,"*American Journal of Agricultural Economics,* Feb. 67(1):81-86.

Raup, Philip M. 2003. "Disaggregating Farmland Markets," in *Government Policy and Farmland Markets: The Maintenance of Farmer Wealth*, Charles B. Moss and Andrew Schmitz (eds.), Iowa State Press, IA.

Reynolds, John E., and Warren Johnston. 2003. "Micro-Markets for Farmland: The Case of Florida and California," in *Government Policy and Farmland Markets: The Maintenance of Farmer Wealth*, Charles B. Moss and Andrew Schmitz (eds.), Iowa State Press, IA.

Robison, Lindon J., Robert J. Myers and Marcelo E. Siles. 2002. "Social Capital and the Terms of Trade for Farmland," *Applied Economic Perspectives and Policy,* 24(1):44-58.

Roe, T., A. Somwaru, and X. Diao. 2003. "Do Direct Payments Have Intertemporal Effects on U.S. Agriculture?" in *Government Policy and Farmland Markets: The Maintenance of Farmer Wealth*, Charles B. Moss and Andrew Schmitz (eds.), Iowa State Press, IA.

Schnitkey, Gary. 2010. "Farmland Price Outlook: Are Farmland Prices Too High Relative to Returns And Interest Rates?" *Farm Economics Facts & Opinions* FEFO 10-17, University of Illinois at Urbana-Champaign, Oct. 18. http://www.farmdoc.illinois.edu/manage/newsletters/fefo10_17/ fefo 10_17.html

Schmitz, A., and R. Just 2003. "The Economics and Politics of Farmland Values." Pgs 53-79 in C.B. Moss and A. Schmitz, eds. *Government Policy and Farmland Markets: The Maintenance of Farmer Wealth.* Iowa State Press, Ames, Iowa.

Shalit, H., and A. Schmidt. 1982. "Farmland Accumulation and Prices." *American Journal of Agricultural Economics* 81 (Nov):710-19.

Sherrick, B.J. and P.J. Barry. 2003. "Farmland Markets: Historical Perspectives and Contemporary Issues," in *Government Policy and Farmland Markets: The Maintenance of Farmer Wealth*, Charles B. Moss and Andrew Schmitz (eds.), Iowa State Press, IA.

Shi, Yue Jin, Timothy T. Phipps, and Dale Colyer. 1997. "Agricultural Land Values under Urbanizing Influences," *Land Economics*, Feb. 73(1): 90-100.

Shultz, Steven D., and Steven J. Taff. 2004. "Implicit Prices of Wetland Easements in Areas of Production Agriculture," *Land Economics*, Nov., 80(4): 501-12.

Siniavskaia, Natalia. 2008. "The Effect of the Home Building Contraction on State Economies." Special study published by Housing Economics: The Economics Publication for America's Housing Industry. http://www.nahb.org/generic.aspx?sectionID=734&genericContentID=99676&channelID=311

Soule, M.J., A. Tegene, and K.D. Wiebe. 2000. "Land Tenure and the Adoption of Conservation Practices," *American Journal of Agricultural Economics* 82:993-1005.

U.S. Department of Agriculture, Economic Research Service, 2011a. Farm Risk Management: Government Programs and Risk. http://www.ers.usda.gov/Briefing/RiskManagement/GovernmentProgramsandRisk.htm;

U.S. Department of Agriculture, Economic Research Service, 2011b. Farm and Commodity Policy: Program Provisions: Crop Yield and Revenue Insurance. http://www.ers.usda.gov/briefing/farmpolicy/insure.htm

U.S. Department of Agriculture, Economic Research Service, 2011c. Farm and Commodity Policy: Government Payments and the Farm Sector Briefing Room (last updated September 9, 2011). http://www.ers.usda.gov/Briefing/FarmPolicy/gov-pay.htm

U.S. Department of Agriculture, Economic Research Service, 2009. Rural Amenities: Natural Amenities Scale Briefing Room (last updated October 22, 2009). http://www.ers.usda.gov/Briefing/RuralAmenities/naturalamenities.htm

U.S. Department of Agriculture, National Agricultural Statistics Service. 2009a. Agricultural Land Values and Cash Rents Annual Summary. http://usda.mannlib.cornell.edu/MannUsda/viewDocumentInfo.do?documentID=1446.

U.S. Department of Agriculture, National Agricultural Statistics Service. 2009b. The Censuses of Agriculture, 2007, 2002, 1997, and 1992. http://www.agcensus.usda.gov/index.asp

U.S. Department of Agriculture, National Agricultural Statistics Service. 1999. Agriculture Economics and Land Ownership Survey. http://www.agcensus.usda.gov/Publications/1997/Agricultural_Economics_and_Land_Ownership/index.asp

USDOT-Bureau of Transportation Statistics, 2004. National Transportation Atlas Database. Research and Innovative Technology Administration.

Wunderlich, G. 1997. "Land Taxes in Agriculture: Preferential Rate and Assessment Effects," *American Journal of Economics and Sociology,* April, 56(2):215-28.

Xu, F., R.C. Mittelhammer, and P.W. Barkley. 1993. "Measuring the Contributions of Site Characteristics to the Value of Agricultural Land," *Land Economics*, Nov., 69(4):356-69.

Zulauf, C. 2007. "Fixed Payments and the Major Farm Program Crops: An Overview." AEDE-RP-0077-07, The Ohio State University, Department of Agricultural, Environmental, and Development Economics.

APPENDIX A: DATA INPUTS

Euclidean Distance Metrics

Distances to three types of features were calculated using a Euclidean distance (straight-line distance) tool in ArcGIS. The output cell size for these Euclidean distance grids was 1 mile.

Interstates

The source of the interstate data is the US National Transportation Atlas Interstate Highways, through ESRI, from 2002. A one-mile resolution Euclidean distance grid was calculated from interstates out to a maximum distance of 2,000 kilometers.

Grain elevators

The best available set of terminal and subterminal grain elevators are derived from Baking Business Magazine/Souza Publishing (2002). A 1-mile resolution Euclidean distance grid was calculated from grain elevators out to a maximum distance of 2,000 kilometers. This set excludes country grain elevators.

Water ports

The port data contain physical information on commercial facilities at the principal U.S. Coastal, Great Lakes, and Inland Ports (USDOT

Bureau of Transportation Statistics 2004). The data consist of listings of a port area's waterfront facilities, including information on berthing, cranes, transit sheds, grain elevators, marine repair plants, fleeting areas, and docking/storage facilities. Collection of data is performed on a rotational basis to ensure onsite accuracy at each facility. A 1-mile resolution Euclidean distance grid was calculated from water ports out to a maximum distance of 2,000 kilometers.

Counties

Counties come from the Environmental Systems Research Institute counties from the year 2000. The county polygons were converted to a 30-meter resolution grid.

1099 Tax Information

Totals of FSA program payments by county were summarized and extracted from the FSA database for 1999-2008. These were joined to counties, and converted to a 1-mile resolution grid for each year.

Indemnity Payment Data

Totals of RMA indemnity payments by county were summarized for 1999-2008 and converted to 1-mile resolution grids.

Ethanol Processing Capacity Surfaces

Using the ethanol capacity attribute of ethanol plants, ERS generated a grid using the ArcView's kernel density function tool. The output cell size is 4 kilometers, and the capacity is distributed over a 125-kilometer distance from each plant. This study considers the ethanol capacity grids from 2002 through 2008. The kernel density approach is a way to demonstrate the expected decay of influence from ethanol plants on land value as a function of distance.

Land Capability Class (LCC)

This is one of the attributes available in the Natural Resources Inventory (NRI). Its purpose is to classify soils into one of eight classes describing suitability for agricultural production. The county-level measure used in this study is the percent of the county cropland area composed of LCC classes 1 and 2, the two best suitability classes. Inherent soil qualities—such as wetness, erosivity, stoniness, shallowness, and drought properties—are considered in the class assignment; however, anthropogenic soil enhancements are not considered.

For the official NRCS definition of LCC, visit http://www.nrcs.usda.gov/technical/NRI/1997/summary_report/glossary.html.

For LCC class and subclass descriptions, visit http://www.nrcs.usda.gov/technical/NRI/maps/meta/m6175.html.

Crop Yields

Average corn yields were derived from the NASS Crop County Estimates database. Yields were averaged from 2001 to 2006 at the county level to get an overall yield expectation.

PIZA

The ERS Population Interaction Indexes to Classify Agricultural Land (PIZA) is a measure of urban influence on agricultural lands available in a 5-kilometer resolution grid.

See Appendix B and http://www.ers.usda.gov/Data/PopulationInteractionZones/accessdata.htm

Population Interaction Index (PII)

The Population Interaction Index (PII) is designed to provide a cardinal measure of the potential interaction between nearby urban-related population and agricultural production activities in each grid cell. This is a 5- kilometer resolution grid.

See Appendix B and http://www.ers.usda.gov/Data/ PopulationInteractionZones/accessdata.htm

Natural Amenities Scale

The ERS natural amenities scale is a measure of the physical characteristics of a county area that enhance the location as a place to live. The scale was constructed by combining six measures of climate, topography, and water area that reflect environmental qualities most people prefer. These measures are warm winter, winter sun, temperate summer, low summer humidity, topographic variation, and water area. The data are available for counties in the lower 48 States.

The file contains the original measures and standardized scores for each county, as well as the amenities scale. This county-level measure was converted to a 1-mile resolution grid. See http://www.ers.usda.gov/Data/NaturalAmenities/

Appendix B. Alternative Measures of Urban Influence

Population Interaction Index (PII)

The first alternative measure of influence is the population interaction index (fig. B1). The population interaction index ranks each location or area on a continuous scale in terms of the potential interaction between nearby urban-related population and agricultural production activities. The index is created for a series of 5-kilometer grid cells laid out across the contiguous 48 States. The index for a single location takes the form:

$$PII_i = \sum_{j=1}^{N} \frac{P_j}{D_{ij}}$$

where P_j is the population of cell j, and D_{ij} is the distance between cell i and cell j. The measure is aggregated across N grid cells within a 50-mile radius of cell i. The index increases as population increases and/or as distance to the population decreases.

PII therefore provides a measure of urban influence that accounts for differences in population size of urban centers, as well as the potential for multiple nearby urban centers. Figure B1 demonstrates that the average value per acre for farmland declines as urban influence decreases. The decline is greatest at the highest levels of PII or the greatest amount of urban/rural interaction. At the greatest level of population interaction, the average value per acre is $9,515; at the lowest level, the value is $2,252.

Population-Interaction Zones for Agriculture (PIZA)

The population-interaction zones for agriculture (PIZA) identify agricultural land use in which urban-related residential, commercial, or industrial activities affect the economic and social environment of agriculture. Building on the PII grid cell data, PIZA classify grid cells into either rural or population interaction zones based on threshold values. The rural zone includes population that supports active commercial farming, including related input and output industries. The threshold values vary for each of the 20

USDA Land Resource Regions to account for differences in farmland productivity. Grid cells initially classified into the population-interaction zone were further classified into one of three categories representing increasingly higher levels of population interaction.

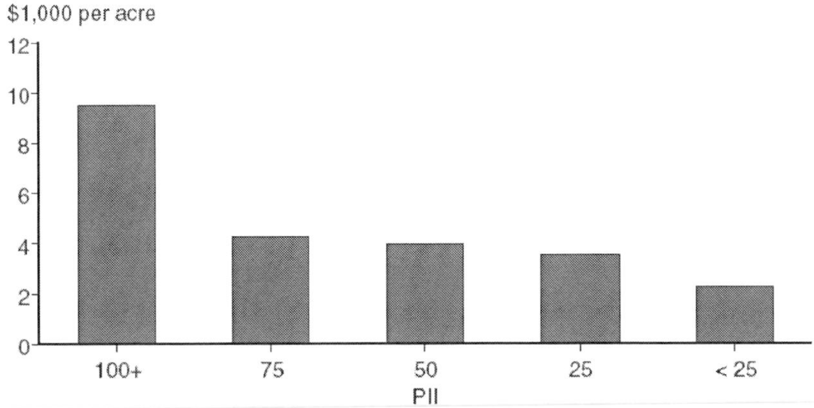

Sources: USDA, ERS analysis of National Agricultural Statistics Service June Area Survey data and PIZA data.

Figure B-1. Farmland value per acre by PII.

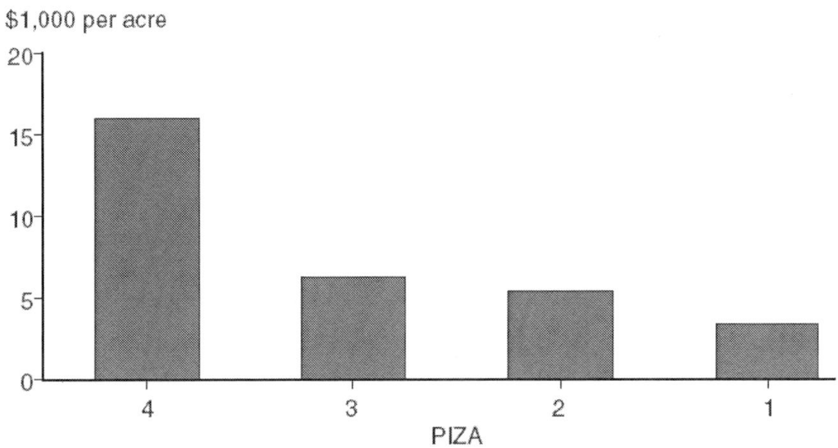

Sources: USDA, ERS analysis of National Agricultural Statistics Service June Area Survey data and PIZA data.

Figure B-2. Farmland value per acre by PIZA.

PIZA classifies each grid cell into one of four categories:

- 1 = rural
- 2 = low population interaction
- 3 = medium population interaction
- 4 = high population interaction.

The figure below depicts the average value per acre for all agricultural lands by PIZA classification. Similar to the previous measures of urban influence, the value per acre is highest in areas with the greatest degree of population interaction and declines as population interaction decreases. For areas with high population interaction, the mean farmland value per acre is $16,041, versus $3,408 in rural areas.

End Notes for Performance of Farmland as an Investment

[1] Numerous studies have examined the applicability of the present value model to farmland prices (e.g., Falk, 1991; Clark et al., 1993) and considered other factors that can affect farmland values, such as inflation and risk relative to nonfarm investments (Chavas and Thomas, 1999; Irwin et al., 1988) and local (parcel-specific) factors discussed later in this report, such as amenities and the value of land in a developed use (e.g., Libby and Irwin, 2003; Livanis et al., 2006).

[2] Net farm income in figure 4 is that for the farm sector as a whole. It is computed by determining the gross value of goods and services generated by farm assets and subtracting the expenses incurred during the year. All cash and noncash items are included in gross farm income and in expenses, regardless of whether they were earned or incurred by farm operators, contractors, or farmland owners.

[3] The farm sector includes about 800,000 farms where the operator identifies farming as the primary occupation and 1.2 million farms where the operator is not engaged in farming as a primary occupation. As a result, the latter group does not generate as much income from farming. For this group there is a substantial gap between farm income and the value of land and buildings on their operations. In contrast, the remaining 800,000 farm businesses that do generate substantial income from farming have more consistency between land/building values and farm earnings.

[4] The relationship between land rents and values was a popular research topic during the run-up and later collapse of land prices during the 1970s and 1980s. A central question of this research was whether a simple capitalization formula (value=[annual rent / capitalization rate]) could predict land prices in the current year based on current rental rates, or whether a more sophisticated model is needed (e.g., Melichar 1979, Barry 1980, Alston 1986, Burt 1986). Rents were found to help explain land price variation in some parts of the country, while in other areas the increasing influence of non-agricultural factors required more sophisticated modeling approaches (e.g., Featherstone and Baker 1987).

[5] More farmers in these three regions use irrigation, and some of the volatility could be due to our averaging of irrigated and non-irrigated rents and land values.

[6] Although some Corn Belt States like Ohio have extensive areas of low-density urban influence, ERS estimates that nearly three-quarters of farmland in the Corn Belt is not subject to urban influence.

[7] The amount of cash available to service debt on the purchase of all land and buildings is calculated as revenues from all crops and livestock produced, government payments, insurance payments, and cooperative dividends less operating and ownership expenses (i.e., adjusted net returns). To determine the maximum value of real estate that can be purchased from the adjusted net returns, we assume a loan term of 30 years and use average annual mortgage rates for this type of loan. This is then used to calculate the maximum real estate value that could be mortgaged based on agricultural production only. Trends over time in farm real estate affordability can be examined using data from the farm sector accounts (http://www.ers.usda.gov/Briefing/ FarmIncome/).

[8] Nonfarm factors include purchases for recreation uses or for residential or other types of development.

[9] Estimated using USDA-NASS data from the 1999 Agricultural and Land Economics Survey (Sherrick and Barry 2003).

[10] These estimates are based on USDANASS/ERS' Agricultural Resource Management Survey of farm operators, and account for about 71 percent of land in farms. Estimates are not available on how land was acquired by non-operating landlords, who owned 29 percent of farmland in 2007. Non-operating landlords own farmland and rent it to farmers rather than operate it themselves.

[11] This is consistent with the low affordability findings noted in figure 9.

[12] The hedonic technique—a statistical modeling approach economists often use when modeling the determinants of farmland values—assumes a parcel can be identified by a unique set of attributes and that the value of a land parcel is an aggregation of the values of its individual attributes. It is an indirect approach that allows researchers to substitute proxies for actual measures of net returns which are typically not available (Freeman, 1993).

[13] Results were similar when the population size of cities was set at 25,000.

[14] We examined the correlations using different years to test the sensitivity of the findings, but the patterns did not vary across years.

[15] Other non-land inputs would also likely capture some of the benefits of government payments when farmers respond by increasing production.

[16] Payments are made to landowners if they operate the land. If the land is rented, the payments accrue to cash renters. Under a share-rent agreement, payments are split between the land-owner and share renters. Landowners who rent out the land and don't receive the payment directly may capture the benefits of the income stream through bidding up the lease payments.

[17] Administrative data reported for tax purposes (1099 tax information) were used to calculate direct payments. These data did not disaggregate direct and countercyclical payments, but in 2007 the total expenditure through the DCP for direct payments was approximately $5.1 billion versus $1.1 billion in countercyclical payments. DCP payments from the 1099 tax information, and RMA indemnity payments by county, were converted to 1-mile resolution grids and combined with the JAS sample points. See Appendix A.

[18] Counties with ratios of both direct payments and premium subsidies to land values below 0.01, as well as counties with no JAS observations, appear as white on the map. In addition to excluding counties that are in the bottom 20 percent of cropland acres, counties without at

least nine tract-level observations from 2006 to 2008 are excluded; these also appear as white. This ensures at least three farms were in the county sample from 2006 to 2008, as farms remain in the JAS for 5 years.

[19] The Housing Price Index (HPI) is published quarterly by the Federal Housing Finance Agency (FHFA, previously the Office of Federal Housing Enterprise Oversight) (Calhoun, 1996). HPI is a broad measure of single-family house prices. The measure is a weighted, repeat-sales index drawing from repeated sales or refinancing of the same property whose mortgages have been purchased or securitized by Fannie Mae or Freddie Mac. The HPI includes house price figures at a number of spatial intervals, including Census division, State, and metropolitan statistical area (MSA). The nonmetropolitan HPI measure (which we call RHPI) has been published at the State level by FHFA quarterly since 1995. The measure includes all repeat sales mortgages in nonmetro areas as defined by the Office of Management and Budget (OMB).

[20] Easements are also used to protect land from conversion to cropping uses. For example, Shultz and Taff (2004) found that easements to protect wetlands significantly lowered land values in North Dakota.

[21] Also, in 1999, over 80 percent of landlords leased to only one tenant (USDA-NASS, 1999).

[22] One study estimated that of the $40 billion in total land value attributable to farm program payments in the Heartland, $25 billion is related to land owned by non-operators (Barnard et al., 2001).

[23] USDA's Conservation Reserve Program (CRP) and Wetlands Reserve Program (WRP) are two conservation programs that retire cropland from production. While non-operators might be more reluctant to participate due to contract lengths (at least 10 years in CRP, and 30 years in WRP), the benefits of enrolling land include reduced farm labor requirements and payments to the landowner. However, it is not always clear whether landlords are classified as NOL or operators when they have land, particularly whole farms, enrolled in these long-term land retirement programs.

[24] The AELOS does not distinguish towns and cities based on population size, so it is possible some landlords report living in towns that would be classified as "rural" according to Census Bureau definitions of urban areas and clusters.

INDEX

G

H

N

O

P

T

U

V

W

Y